CASSIOPEIA'S QUEST

A Young Woman's Journey of Exploration, Learning, and Imagination.

Hers is an odyssey that will take her through some of the events in history that inform and illuminate our humanity.

Jerry Smetzer,

Writer and Painter

Copyrights.

This is a work of historic fiction. There are many fictional characters in each story who are products of my imagination. They bear no relationship to any real person either alive or dead. The fictional characters interact among themselves and among other named characters who are said to have existed in history or who have been referenced in the record of historic facts, figures and events.

Names: Smetzer, Jerry, author.
Title: Cassiopeia's Quest : a collection of short works of historical fiction by Jerry Smetzer.

First Edition, Second Printing

For all the children and grandchildren

PREAMBLE

The concept that is driving the creation of this book began in my mind, as I think about it now, several years after seeing the 1971 Australian movie *Walkabout*. The movie is taken from an award-winning book of the same name published in 1959 by James Vance Marshall. The book was re-released by the *New York Review of Books* in 2012.

The Movie's Director, Nicholas Roeg, lays the opening scenes in a modern Australia city, probably Sidney. We are introduced to an upscale suburban white family. There is a man in a tie, probably the father, who seems troubled. There is a post-puberty private school girl and her much younger, sun-bleached blond brother. Both children are dressed in the pressed uniform and hat for their school.

While these early scenes unfold a young Aboriginal boy begins his tribal rite of passage deep in the outback. This is his walkabout - the concept behind the movie's title. He will have a bow, some arrows, a knife, a possibles pouch of tools, charms, and totems for the unexpected events that may challenge him during his journey of weeks, perhaps months, across the parched and trackless desert.

The essential story begins as the father drives deep into the outback with his children for what might be a picnic. Suddenly, an event too terrible to discuss here leaves his two children without their father. In an instant they are abandoned, traumatized, and frantic.

After a few days of aimless wandering, desperate to find water, their school clothes dirty and ripped, their paths cross with that of the black, nearly naked Aboriginal. The Aboriginal knows how to find water, but he speaks no English, and the cut-up bloody parts of a large lizard hanging from the decorated piece of woven cloth that passes as a waist band is gross and disgusting to the high-born children.

He has no experience of the pale and frightened children before him, nor of their parentage. Their strange dress tells him nothing of their tribe. He cannot understand why they are reluctant even to touch the few mouthfuls of muddy but drinkable water the

Aboriginal has found in a damp spot at the bottom of a shallow depression.

This collision and merging of two vastly different cultures, in all their gritty, grisly, and conflicting detail, has become a powerful metaphor, for me, for all cultures in all times and in all places. The seeds of this concept began to germinate during my service in the U.S. Peace Corps in the late 1960s. I was the only non-Afghan member of an Afghan land survey crew deep in the deserts of Afghanistan. Roeg's movie gave my concept of clashing cultures the kind of clarity and focus needed to sustain a book writing project of this size and scope.

In *Cassiopeia's Quest* the churn of merging cultures is the binding force that holds these 10 independent stories together over the 100,000 years of their chronology.

The rest is technique. I am a constant reader and regular movie-goer. The structure, visuals, and story-telling in books and movies has guided my thinking - and my aesthetic - throughout the year-long writing of this book. The other part of the skill set required is the actual writing. Hopefully, my Bachelor of Arts Degree in English many, many years ago will now help me carry that load. The "Personal Note" and the "Acknowledgements" sections in the back pages, have more on these topics.

Cassiopeia's Map of theWorld

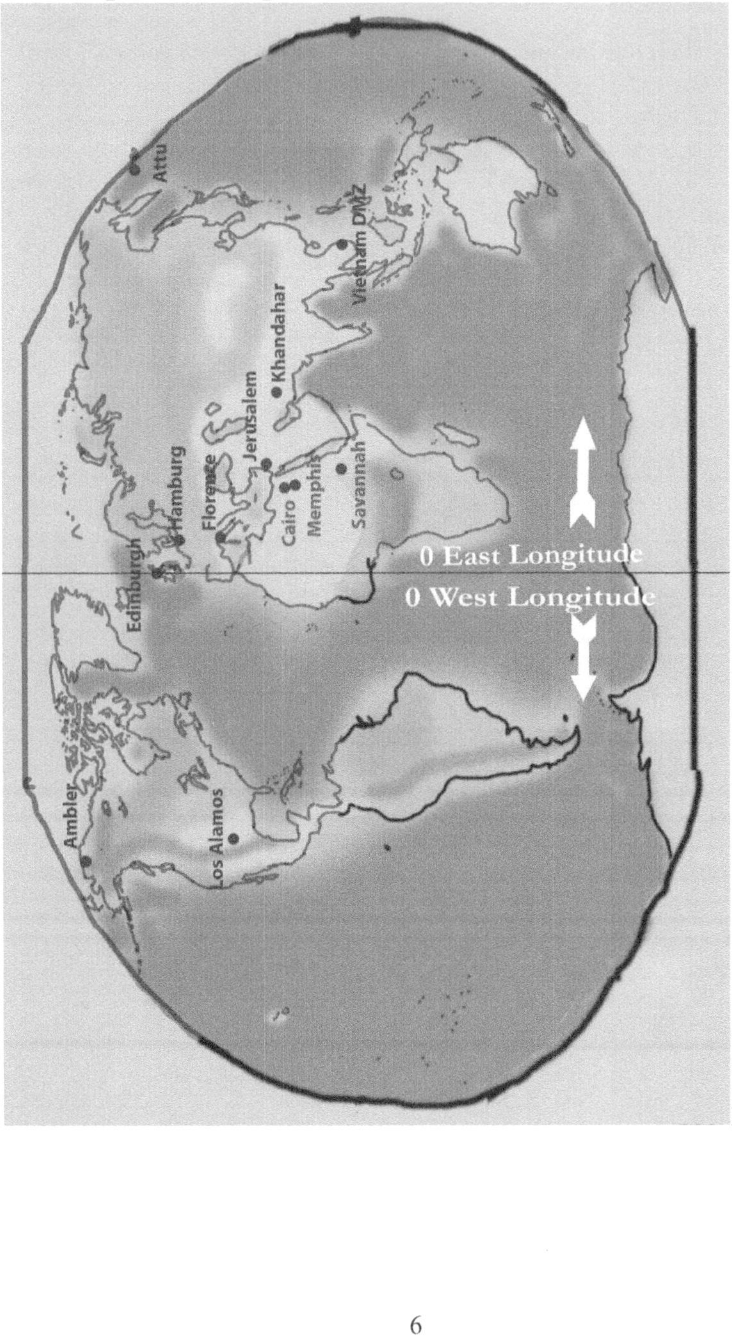

TABLE OF CONTENTS

CASSIE'S STORY

I'm adopted. My mom and daddy named me Cassie as soon as they had all the paperwork done so they could take me home from the hospital where I was staying with my real mom when she died. I was only 3 years old and do not remember much of my first meeting with my parents. All I know for sure is that my mom and daddy have always loved me, cared for me, and taken me into their home as though I am their own child.

Cassie is short for Cassiopeia. Cassiopeia is a constellation visible in the night sky a few degrees on the other side of the North star opposite the Big Dipper. My namesake constellation is not as easy to recognize as is the Dipper, but my position opposite the Dipper, across Polaris, makes me an important object in the night sky to those who must keep the time, or navigate from place to place.

I know I am important, but I do sometimes wish I was as pretty and as easy to recognize as the Belt of Orion in the southern sky.

We live on a farm in northern Ohio where we have some animals, and we grow some corn, wheat and soy beans. There is a small farm house, a barn for storing feed, and a workshop and sheds for the tractor and corn picker. I am 15 years old, and I am scared that I might grow up before I am ready. My daddy helps work the farm, but mostly he is a mechanic who can fix anything that is broken. He is very tall. His hands are usually dirty. Even after washing them his fingernails still have dirt under them. During the growing season he is gone a lot because he goes around to farms in the area fixing farm equipment.

I love it when daddy takes me with him when he is called to work on a piece of farm equipment, especially when the broken equipment is holding up the harvest, or the spring plowing and planting. The concerns of the farmers who must get their equipment back into operation sometimes means that daddy calls on me to hand tools to him, or to hold some piece of the equipment in place while he tightens a bolt, installs a carburetor, or some such. I love being part of the excitement, doing useful work.

My mom is pretty. She likes to wear a pretty ribbon in her hair even when there is no one to see it. She mostly stays at home doing the cooking and cleaning, feeding the animals and taking care of things around the place when daddy is not here. Mom never went beyond 8th grade, but she has many books of her own, and is always delighted when one of her quilting friends bring around a new book. She has read with me every day since I can remember. She doesn't much do that anymore because I learned early how to read by myself. Besides now mom asks me to help her select colors and a design pattern for a quilt she wants to make. I enjoy that so much more than being read to.

My daddy reads too, but he also taught me to shoot and hunt. Now I go out to the pasture back behind the barn and shoot whenever daddy gives me some bullets. He gave me a 22 bolt action rifle when I was 8. He looked at me serious, then, and told me he wanted me to learn how to help feed the family.

I do try really hard to help with chores. Mom and daddy keep us all well fed with home-grown chickens, a small garden, a couple of hogs, and milk from a scrawny goat daddy named Jack Rabbit. I like working with all the animals, and sometimes give them names. In the fall, though, we slaughter the chickens. Daddy takes the heads off the chickens with an ax, I dunk their bodies in hot water, then pull off the feathers. Mom cuts them up and wraps them so they last through the winter. After helping to raise them, I don't like the slaughtering, but I know it has to be done if we are to have decent food to eat.

Mom and daddy both say I talk too much, especially when we sometimes have guests over to the house. I have always talked a lot, and I ask lots of questions. I don't have any fear of treating adults like I was as old as they are. When they say something dumb I make a face. I can't help it. I didn't really care if they told my parents behind my back that I was rude child. Maybe my parents should have been strict with me so I would be more polite with the neighbors. But they never were. Mom would sometimes say: "Cassie please settle down.' Once, daddy told me to shut up, but there was only that one time. After he said it, he seemed very sad for awhile, and couldn't look me in the eye.

Whenever my dad's head dropped in a sad frown after one of my dumb remarks that was always the worst punishment anyway.

When I went to school, I read every book I could find, and sometimes had to hide to do it. My best teacher, Mrs. Hastings, always tried her best to give me assignments and books to read that I would like, rather than those assigned to our class by the Principal of our school. I didn't like those books very much because I had mostly already read them.

I wanted to know about the world; about history; about science; about art. I wanted to know what others thought of these things. I was often blunt in asking older people what they thought about things, even really personal things.

Mom and daddy liked each other well enough, but they had always had to work hard to keep the farm a decent place to live and with enough food to eat. They never really had time to help me understand all the things I was learning about. Their parents before them were raised in the depression and worked especially hard on this farm to make sure there was plenty of food on the table, even if there was never much money for pretty things.

When I was thirteen, my dad bought me a 3-speed bicycle for Christmas. In the summers I was never home after that except when tornados were forecast. There was another kid more or less my age who lived on a nearby farm. Before we both got bikes our parents would get us together for a visit. As soon as we could break free from the parents, we would go out to explore the neighborhood, and whatever farm equipment, barns, or animal pens we could find. Later we biked to our closest little town three miles away so we could hang out with our other school friends. There was a bigger town seven miles beyond, and the bunch of us would sometimes ride our bikes to that town, and marvel at all the stores and tall buildings. In my bike travels I became more independent, more aware, and more curious about the complexity, diversity, and all the wonders of my small farming community, and the larger world beyond. I and one of my more adventuresome friends, Eddie Clark, traveled further, even dangerously further, into places far away from home. These adventures began to give me more focused interests, and about how those interests would fit when I grow up. After seeing the pleasure that adults took in listening to the telling of good stories, I wanted to learn how to tell good stories too.

So, in quiet times, I wrote little stories about myself and my life on the farm. My parents were my audience as I read them aloud.

But I am getting ahead of this story, and this story is the one I really want to tell. The thing I want to talk about is the day my daddy woke me up early on a cold and crisp fall morning and told me to bring my rifle out to the hog pen. I had learned how to use the rifle in the 5 years since daddy bought it for me. I was a pretty good shot when I was shooting at tin cans and bottles. Usually, when daddy gave me some bullets to practice with he would send me on my way with a smile, and a head rub.

This morning, he was very somber. As soon as he was sure I was awake he turned and left the room without speaking. When I went out, I saw my Uncle Sid's pickup truck in the driveway. I remembered then that it was hog butchering day. My daddy called on his brother Sid to come over because Sid – actually Uncle Sid to me - knew all the steps involved with efficiently killing and cutting up a hog for winter meat.

Once the hog was down, it was drug away and stung up by the hamstrings on a pulley rig suspended on tall tripod. Underneath the strung-up hog was a barrel of hot water. The hog was dropped into the hot water, left a few minutes then lifted out. Uncle Sid, Daddy, and Uncle Sid's helper, Showalter would then use a special tool to scrape the bristle off the hog's hide.

I didn't like having Showalter around. In the summer time he wore a floppy hat and bib overalls, but no shirt or underwear. He had looked at me in the garden the previous summer in a way that made me uncomfortable. I had never watched the butchering up close. Usually, I looked out from the kitchen window when I heard the shot that killed the hog. Then I turned back to my kitchen chores. This morning, though, they had not yet killed the hog, and I began to feel chills as daddy motioned me to come out to the pen by the tractor shed. Uncle Sid was another big guy like daddy, but I didn't like him very much either. He was gruff, he swore a lot, and he treated me like I was a child in the way of his important business.

"You gonna let this little shit shoot that hog?" Uncle Sid said. Daddy didn't answer. He motioned me to his side. "Did you clean the rifle, Cassie, and check the rounds before loading?" "Yes, daddy." I said, but my lower lip was starting to quiver. I began to think of the chickens that mom and daddy killed in the fall, about the time I spent feeding them, getting to know them, and sometimes naming them; about how much I hated to think about their being killed so we could eat. I thought I might cry, but daddy put his hand on my shoulder, and squeezed hard. "Let me see your rifle," he said. I lifted it to him. He took it, looked at it, and pointed it toward the sky. He partially cocked the bolt back to check the round, then he closed the bolt, and handed the rifle back to me.

"Do you see that hog over in the corner of the pen with the black spot across his left rump?" Yes, daddy." It was all I could do to keep from bursting into tears. "I want you to walk up to the fence and rest your rifle on that railing so you are comfortable. Then I want you to aim your rifle between his eyes, right here." Then, with the index finger on his right hand, Daddy pointed at a spot between his own two eyes. I could see the dirt under his fingernail. "Then I want you to pull the trigger." I felt frozen to the spot where I was standing. "Go on, now, Cassie. This is something you need to do."

Sid said: "Jesus Christ, Ed (Ed was my daddy's name.) Why don't you shoot the goddamn hog. Yer kid's scared to death, and I got work to do." I looked at daddy. He looked into my eyes, his face a mask. I looked briefly at Sid, then turned and walked toward the pen. I laid my rifle on the rail like daddy said then looked through the iron sights at the hog's forehead.

The hog looked at me. I drew the rifle barrel down to a point right between the hog's eyes.

I know all the animals have a sensibility. They have friends and neighbors among their own kind; they have a language; hogs especially are very social and very intelligent. I saw all those things in that hog's eyes. My only thought was that I had not named either of our hogs. That made them almost strangers to me, but, with my rifle aimed at the hog, and my daddy watching, I could take no comfort in that.

I squeezed the trigger. The hog dropped in place, gave a brief quiver along his flanks as his eyes went out of focus and he died. Uncle Sid said: "About goddamn time." He climbed over the fence with a knife and slit the hog's throat so the blood would run out onto the ground. I felt dizzy. I pulled the rifle close to my body with the barrel skyward, then flopped down on my butt, totally drained… too drained of energy and emotion to even cry or think.

Daddy came over and took the rifle from me and opened the bolt to check that there was no shell in the firing chamber. "Good job," he said, and rubbed my head, then set the rifle up against a tree near me. Uncle Sid went to get his butcher's saws and knives out of his pickup. Daddy and Showalter then strung a length of chain through the hog's rear hamstrings so they could drag it out of the pen, and lift it up to the tripod. I moved out of their way, then sat down again. I had never felt so tired and weak. I watched for awhile as Uncle Sid, Showalter, and daddy stripped the bristle off the hog, then cut the gut open so the intestines and internal organs could spill out. Later I went back into the house.

Showalter's eyes followed me as I walked away. The last words I heard was my daddy speaking harshly to Showalter. "Get your mind back on your business, Showalter, or I am going to kick your ass."

Cassiopeia is a constellation visible in the night sky a few degrees on the other side of the North Star from the Big Dipper. In Greek mythology, Cassiopeia was the Queen to Cepheus, King of Ethiopia. Together, they ruled Ethiopia. Like most Ethiopians, the family was black. There was no racism in my family, but, secretly, I took pleasure in the blackness of Cassiopeia, and in the feeling of an earthy strength and otherness it gave me. Cassiopeia was the mother of Andromeda. Both were proud of their beauty, and both spoke of it to all in their presence.

This pridefulness was a problem for the gods of the day, and particularly for Zeus, the main god in the Greek pantheon, and Poseidon, the Greek God of the Oceans. Poseidon was so angry with it that he resolved to punish both women. Andromeda was stripped naked and chained to a rock in the sea. There she was to become food for the sea monster Cetus. Cassiopeia was banished to the sky, forever tied to her throne, and forever locked into a rotation around the North Star, and, so, upside down to us for some part of each night.

Before I could fully absorb the meaning of all these wonders of myth and cosmos taken from the ancient Greeks, I got diverted from my studies and brought back down to earth, hard, by the bloody curse that reminds me, still, that I am no longer a child. Menstruation, and the expanding changes in my breast, butt, body hair, and sexual awareness that come with it, caused me to wonder about the natural world, about the ancient mysteries of our species, about the roots of our humanity, about the orbits of the moon and planets, and about the evolutions of all things.

PANEL 1 - SAVANNAH

Ah'-oom – a talented hunter and relatively hairless young girl – leads a group of hunter-gatherers along a north-flowing river toward a herd of beasts; at least one of the beasts must be taken for the meat, fat, bone, hide, and sinew needed by her extended family.

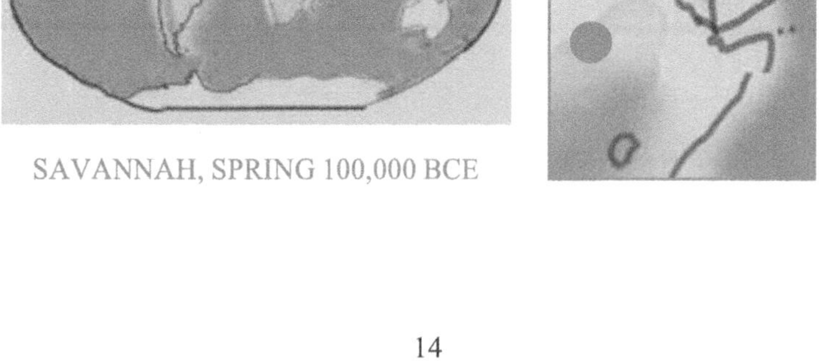

SAVANNAH, SPRING 100,000 BCE

PROTAGONIST

Ah'-oom – A young girl, 2 or 3 years past puberty, who is a talented live-animal tracker, hunter, and hunt leader.

Ar-gah' – A male youth who was found, alone, probably lost from a competing group of hunters many years ago. He is older than Ah'-oom and is jealous of her hunting talent. He often competes with her during the infrequent hunts for kills that might earn for him the favor of the group.

LESSER CHARACTERS

Boa – A senior male and Yah, a female, who are the parents of Ah'-oom.

Yah'-nay – The oldest member of the group. She is a star-gazer, singer, story teller, and healer.

Boa-dee' – An older brother of Ah'-oom; and Pe'-dee, a younger sister of Ah'-oom who gathers wood and edibles, she also collects things that are pretty, or interesting in size, shape, or color.

MINOR CHARACTERS

Bo-nah' – Brother to Boa, and an uncle to Ah'-oom; a hunter and maker of weapons. Bo-nah' has the strongest and most accurate throw of all the hunters.

Bo-nee' – A second uncle to Ah'-oom. Also a hunter and tracker, the most skilled at finding and following animal tracks.

Pe'-dah – An aunt to Ah'-oom, and helper to Boa and Yah'-nay.

Bw'-ak – A raven with a single white feather on top of its right wing. Bw' ak is a guide and spiritual leader to Yah'-nay.

Late in the Pleistocene the ice age continues to extend its grip on much of the northern latitudes of the planet. In the north African savannah it will soon be dawn. Just west of present day Ethiopia, a small group of homo-sapiens hunter-gatherers are huddled together for warmth in the chilly night air.

The scattered bright stars in the dark sky above the African Savannah fade as the sky at the horizon in the east lightens. Behind a mound of earth out of the wind a small group of huddled homo sapiens hunters, bodies touching and arms and legs tangled together for warmth, begin to stir out of their sleep. One, Yah'-nay, the eldest female of the group, is already up and looking upward toward the brightening darkness. Her view of the moon and bright stars in their various patterns has begun to fade in the gathering light.

Yah'-nay looks down toward the position of the sun on the horizon where it will soon rise. She sees that the sun, at the point of rising, moves further north each day. She remembers that the shadows cast by the sun at midday continue to shorten. Yah'-nay knows from this that the rains are passing and the days and nights will be warmer as the sun grows hotter.

She is joined by Ah'-oom, a young woman two seasons past the start of her bleeding. Ah'-oom squats near the older woman and pulls the animal hide garment aside from Yah'-nays shoulder. A purr deep in her chest begins as she picks the tiny vermin buried deep in Yah'-nays body fur. Several of the hunters have fur over most of their bodies, but Ah'-oom has large patches of bare dark skin on her back, chest, and upper arms and legs. She is more prone to discomfort during periods of cold than are the other members of her group. She often needs to wrap herself in extra animal hides for warmth. Yah'-nay is careful to make sure that Ah'-oom has enough furs around her at sleeping time.

Ah'-oom and Yah'-nay take pleasure each in the company of the other. Yah'-nay exhales sharply through pursed lips and points toward a smudge of stars that show clouds gathering in the northern sky. She lifts her nose and sniffs. Ah'-oom also sniffs, then motions for silence. She becomes alert. She stands and looks toward the source of a gathering breeze.

Ah'-oom points to the source of the breeze, and sniffs again. With hand movements and a guttural sound deep in her chest she indicates that several large animals are perhaps a day's distance away. She turns to grab her spear shaft and the bone she has sharpened that will be attached to it. She has notched the shaft to accept the bone along the grooves carved on either side, and she will wrap shaft and bone with a strip of hide leaving most of the spear point and cutting edge clear. She returns to the group and kicks some of the others. She begins to pull three of them away from the other tangled bodies. For those who object, or resist, Ah'-oom gives them another kick, or a sharp punch to the head.

Boa, the senior male in the group gives a low bark toward Ah'-oom. She looks up, and stares at Boa. Boa passes his stone ax, the symbol of his leadership, to her. He motions her and the others toward the source of the freshening breeze, and to the several ravens flying in the same direction. Ah'-oom calls to three of the hunters and shows them the direction Boa has shown her. She also points to the birds.

One of the hunters, Ar-gah', is an adopted member of the group who was probably two seasons older than Ah'-oom. He was first taken in as an injured youth who had been abandoned by his familial hunting group. Boa was impressed that he had found water, had hidden himself from predators, and had survived for several days even though injured, and not able to move quickly.

Had Boa not seen value in the foundling, he may have left Ar-gah' on his own, reliant on his own resources for continued survival. No other member of the hunting group would have argued against it. As it turned out, Ar-gah' has grown into a good hunter, and his loyalty to Boa, and to the welfare of the group is total.

Ar-gah' senses the affection that Boa feels toward him but is unhappy that Boa has not yet felt enough confidence in his hunting skills to give him leadership of the hunting party.

The other two hunters were brothers of Boa's mate, Yah. Bo-nee' is a skilled tracker, able to find the tiniest and lightest footprints in the loose sand and grasses typical of the Savannah away from water, and in the twisted branches and crushed leaves in more shrubby areas. From the tracks, he can then detect the directions of travel. Bo-nah' is the strongest of the group, able to throw a spear a long distance with deadly force and accuracy. He is also very imaginative with the materials used for tools and weapons. As he puts a tool or weapon together, he often pauses to look at the materials, and think about how to improve his throwing ability. Bo-nah' knows that his value to the tribe lies in how well he can inflict a crippling injury from a long distance on the animals the group needs for food, warmth, and shelter.

To the extent possible, Bo-nah' has passed his skills along to Ah'-oom both because Boa has expressed his pleasure in it, and, too, because Bo-nah' takes pleasure in passing his skills along to such a bright, attentive, and capable pupil.

Today, Bo-nah' picks up his spear, and, with it, a long piece of thin leg bone he has been working at one end with abrasive stone. He has abraded a pocket in the end of the bone in a way that can hold the back end of a spear shaft. He has not yet tried out his idea in a hunt, but the parts fit well together, and it appears it will, with practice, make the force of his throw more powerful, and the accuracy more true. He hopes he will have an opportunity to use the bone with the spear pocket, and the spear, in the course of the hunt.

Ah'-oom and the hunters move down to the creek that runs near their sleeping place. She urges them to drink as much water as they can stand and carry as much as they can in their skin bags. They stop to pick up their clubs, spears, and strips of sharpened bone, then start moving in the direction Boa has indicated. Ah'-oom picks up her father's ax and follows them.

The sky is light now, and only the brightest stars remain visible enough to be seen. The sun has not yet fully risen, but Ah'-oom, as leader, knows that the day will be hot, and the distance to the game long. Travel will be over sandy, wind-blown desert and grasslands where water may be hard to find. She will have to stay alert for the smell of damp ground if the day turns too hot. Each of the hunters also knows that they must call a break from the hunt whenever small animals, reptiles, insect mounds, or clumps of edible plants are found that must be gathered and shared to feed themselves.

The hunters begin to jog at a pace that all members can maintain through the day. When the sun rises well above the horizon, Ah'-oom runs ahead to the top of a small hill and signals a stop. She shakes her arm in a way that signals her wish for absolute quiet. She sniffs, she twists and turns her body as she tilts her head from side to side, and nods up and down listening for sounds, and sniffing for smells of the animals they are hunting. She looks across the horizon in the direction of their prey for any disturbance among birds, or any animal activity that stirs unusual amounts of dust.

Then she stops, still, and stares, squinting, to the north, generally in the direction her shadow falls as it is cast by the southern sun. Before leaving the mound, Ah'-oom looks back toward the direction they have come. Since leaving their sleeping place, Ah'-oom has been memorizing passing landmarks by observing them as she approaches, then looking back at them after she has passed. She unrolls a piece of animal hide with marks showing the major landmarks others have marked from earlier hunts in the area she and the hunters are traversing.

She looks at the sun's course across the sky, and at the length and direction of shadows cast by the hunter's bodies. She feels, sniffs, and tastes the wind with her tongue, and looks at the movement of clouds. She looks back into the distance for the shapes around the group's sleeping place, and particularly to the Acacia tree standing near the wind-break that marks it. She will need to find and follow these shapes on their return from the hunt if she is to find the place where her family sleeps.

Finally, she descends from the mound and begins jogging to the north. The other hunters follow, matching her pace. As they run within a few yards of each other, Bo-nah's face re-shapes itself and his mouth into a wide grin. He vents a high-pitched, warbling cry of enthusiasm for the chase. The others also cry out. Ah'-oom looks into the distance for the next landmark to follow on their way north.

***

Back in the sleeping place Yah'-nay looks after the hunters as they depart. As the sun continues its rise, and the stars above become harder to see she moves into the shadow of the Acacia tree, she turns to other chores. She picks up a small piece of bone that had been taken from an animal's forepaw.

She has sharpened it to a point by rolling it from side to side against a rock with a rough surface. She picks up some thin strips of leather that she has cut out of an old piece of Jackal hide with a piece of flaked stone. She then picks up two whole Jackal hides. These need to be sewn together so they will be big enough to fit Boa, her son.

Ah'-oom's younger sister, Pe'-dee, disentangles herself from the remaining sleepers, and moves drowsily toward Yah'-nay. Pe'-dee brings a piece of bright pebble that she has found in the creek bed. She offers it to the older woman. Yah'-nay knows about the pebble but has not yet figured out how to use it. Even so, she takes it, and gives a hug to the young girl.

Pe'-dee moves away to begin gathering branches and wood needed to build a fire for warmth and cooking. She has already built a fire place surrounded by rocks. Her fire-starting materials include some shredded fungal material that has been dried in the sun, and two small rocks that, when struck together, produce sparks. After learning from her time with Yah'-nay, Pe'-dee knows she will have to build a fire structure in the fire place by placing the small branches, and larger pieces of wood on top of the burning fungal material in ways that will have good air circulation.

Once she has started a fire, Pe'-dee will go further from camp to begin looking for the insects, lizards, snakes, birds, and bird's eggs needed for the group's food today. With luck, she will also find the more useful small animals that she can track down and kill for some bites of more nourishing food. After bringing small animals back to the older women to be stripped and picked apart for the meat, bone and gristle needed for other tribal projects, the hide and bones will be returned to Pe'-dee, along with other found objects, for her to use on her own projects.

Pe'-dah, the aunt to Ah'-oom, and Yah, mother to Ah'-oom, rouse themselves and come over to the Acacia to sit with Yah'-nay. They will sit for a moment then gather some stones for the hunters to use with their slings. The women will hunt for small game that can be cooked for food, or for plants that could either be eaten, or their stalks separated into thinner strips to be used for tying the animal hides in clothing and containers. This was preferred to the use of more of the scarce animal hides. The hides are better used for weapons, and for tying the bones and gathered wood pieces together for shelter structures and other needed objects.

Boa, usually the group's lead hunter, returned to sleep after the other hunters departed for the hunt. Boa would have accompanied the hunting party as leader, but he has injured his leg. It was too sore to enable him to keep up with the younger hunters on such a long-distance hunt. Eventually Boa rouses himself and joins the women to help them with their tasks. Yah makes a poultice of herbs and mud and applies it to Boa's injury to help ease his pain. Boa touches Yah while she does this to express his pleasure at this gesture, and her touch. Boa and Yah are both concerned that the injury is not healing as quickly as it should. Yah resolves to try some other herbal combinations. She will work with Yah'-nay to find a better medicine.

Ah'-oom has called the hunters to a stop. She has, once again, ascended a small rise to look, sniff, hear, feel, and taste the wind. The sun is now past the peak of its travel through the sky, and they have not yet spotted the game that Ah'-oom has said is ahead of them. All check their weapons and make small adjustments as though getting ready for the hunt itself, rather than just the possibility of a hunt sometime in the future.

Ar-gah', - the adopted one, and the third male hunter - stretches, yawns and scratches his torso as though he is losing interest in the main purpose of the hunt, and may have a different idea. Bo-nah' recognizes Ar-gah''s fading commitment, and wonders what it might mean. Bo-nah' knows that Ar-gah' is a season or two older than Ah'-oom, and somewhat more muscular, though he does not run as fast, cannot throw as accurately, and his trail sense is not nearly as well developed.

More important than that, Boa has given the leadership of the hunt to Ah' - oom. The implied message to Bo-nah is that he is to watch over Ah'-oom to make sure to protect her if her leadership is challenged, and to make sure to bring her back to the sleeping place if she becomes injured. Bo-nee' is loyal to Bo-nah, and will support whatever he chooses to do, knowing that Bo-nah' wants to honor the wishes of Boa.

Even so, Bo-nee' likes the impulsive Ar-gah', and the two are often seen together in the distance while Ar-gah' learns the skills of tracking from Bo-nee'.

Suddenly, Ah'-oom lets out a low whistle followed by a series of chirps. She is mimicking the cry of a bird common in this area, but the three hunters know that this call means that she wants Bo-nee' to join her on the rise. He jumps up with his spear and trots toward Ah'-oom. Ah'-oom points to the ground in front of the rise, and to the copse of trees that lies a ways beyond. Bo-nee' sees slight disturbances in the soil in front of them, and a line of shadows in the trees as though something has passed this way and gone into the woods. He nods to Ah'-oom.

Ah'-oom then turns to the remaining hunters, touches her tongue to the roof of her mouth, and gives a series of five clicks, a pause, then five clicks again. It is a message that they need to approach her, but silently. As they gather around her, she motions Bo-nee' to move off the mound tracking the signs in the dirt and vegetation toward the shadow in the trees. She then motions Bo-nah and Ar-gah'' to follow as quietly as they can. All are crouched low. Ah'-oom waits, then moves off in direction to the right and upwind of the other hunters. She checks the ax kept in the hide belt around her waist, then looks, once again, at how she should hold her spear.

None of the spears used by the hunters are perfectly straight. Even so, with practice and Bo-nah''s help each has learned how to hold and throw an irregular spear in such a way that it hits the target with as much penetrating force as each hunter can bring to the throw. The purpose of the spear in the hunt is to try for a killing strike that will cause massive bleeding. Failing that, the hunter needs to be close enough to the animal, or bird, to cripple, in some way, it's ability to outrun the hunters and escape. If the animal is close enough to strike without the need to throw his spear, then the hunter may be able to shove his spear, or other killing instrument into one of the animal's killing zones with great accuracy.

Failing an immediate kill, the hunter may be able to pull his spear and try again. Such opportunities are rare, however, as are the hunters with enough calm, skill, and experience with the spear to know exactly how to place a killing blow in a particular animal, especially when that animal is charging the hunter, or threatening him, or her, with a snarling growl and bared teeth.

Such an animal, though not yet snarling or charging, suddenly appears and faces Bo-nee' from the edge of the woods. The wind was blowing toward the hunters from the woods so the animal could not yet smell or hear them, but it could see shapes in the distance that did not match the background. Those shapes, the hunters, had stopped moving on Bo-nee's signal. They were now waiting silently and without movement of any kind to see if the animal would begin to move in a direction that gave the hunters a real opportunity to take him.

Ah'-oom, several strides away from the others, had also spotted the animal. She also waited silently and quietly to see what the animal would do next. The edge of the woods where the animal now stood was about as far away as any one of the hunters, even Bo-nah', could possible throw a spear. At such a distance, accuracy would not be much of a consideration. The most likely result of such an attempt would be a scared animal that runs back into the woods. As that thought suddenly crosses her mind, Ah'-oom jumps, screeches, and begins running toward the other hunters. The animal perked up at this disturbance, and then leapt toward the running girl. The animal moved quickly.

Bo-nah', Bo-nee', and Ar-gah' remained crouched, but they gripped their spears more tightly as they each estimated the shrinking distance to the now charging animal. To assure that their spears would have maximum crippling power, they knew from Bo-nah's training that the distance would have to shrink to less than 20 strides before they could risk throwing a spear with any killing or crippling potential. Otherwise, they might lose the spear, the speed of the animal's charge would be undiminished, and their tools for defense would be gone.

Ah'-oom ran quickly toward the other hunters while carefully watching the closing distance between the animal and themselves. Suddenly she jerked to a stop, turned, and began running as fast as she could away from the hunters. The animal was hot on her trail and turned with her, thus presenting the crouching hunters with the full length of its body.

Ar-gah', the impulsive one, was the first to leap up and throw. His spear hit the animal near its' hindquarters but glanced off leaving only a minor cut that would not slow the animal down and would not bleed enough over the next several days to enable the hunters to catch up with it. Bo-nah', the strongest and most experienced, threw next. His spear flew true to the animal's chest but did not pierce the heart. Even so, the spear pierced deeply enough in the lung to cause the animal to stumble in its run and fall tumbling and sliding to the ground.

The three hunters now ran screaming with their weapons to the struggling beast. Ah'-oom, on hearing their yells, looked back, then turned and ran toward the animal with her spear and her ax at the ready. The beast struggled to get up. Still several strides away, Bo-nee' threw his spear in the hopes that he could strike a death blow, but he missed hitting any vital parts.

Bo-nah' now runs at top speed with his stone hammer. He literally flies to the beast and brings the hammer down with all the force he can muster in the hopes of crushing the animal's head. But, in that instant, the animal turns toward the approaching hunters and tries to batt down the object flying at him. The claws catch Bo-nah's throwing arm and deflects the force of his hammer. By this time the three other hunters were on the beast with their sharpened bone knives, tearing into it's vital organs through the soft tissue in the belly.

The beast tries to batt them away, but it's blows are now feeble. Eventually, it's life force drains away, and it lays back to die.

All of the hunters scream in pain and agony. All break down in long sobbing cries at the release of tension from the killing. Bo-nah's arm is injured seriously. All the others have cuts where the animal's blows had caught them in his death struggle. After a few moments of grieving, Ah'oom, herself bleeding from a cut in her scalp and on her hip, nevertheless attends to the others before applying a poultice to her own wounds.

Then, as Ah'-oom applies a poultice to the deep cut on Bo-nee's thigh, he, in turn, twists a piece of hide around Bo-nah's arm to stop the bleeding. Apparently, Bo-nah' suffers no broken bones, but he is woozy and unfocused from loss of blood from his wound.

Ar-gah' was the least injured. Ah'-oom directs him to go retrieve his spear, and then check all their weapons for any damage that needs to be repaired. The beast may have had a mate or been associated with a large pride. Worst of all possibilities, the smell of blood might attract any number of curious beasts – like Jackals - to them. Bo-nay' would need some time to recover from his wounds if he was to be any help in defending them from animal attack and getting them all back to their home territory.

She knew they would have to spend the night away from the sleeping place. Her mind filled with all the problems that would need to be solved if they were all to get back home safely. Most importantly, she wondered how they could bring the hide, the bones, and the meat of the beast that had caused all this grief, back to the others.

She looked toward the woods and contemplated the possibilities for shelter among the trees. Since she and her family have little experience of survival among the trees, she also wonders about the risks of animals sneaking up on them in the night.

Ar-gah' finishes his work on repairing the weapons. Ah'-oom directs him to go into the woods and tell her if there is a safe place to move into for the night. He needs to find some tall trees, some downed trees, and some binding materials like vines. These can be assembled into barriers to protect them and the salvaged parts of their killed animal – now buried to cover the smell of blood and rent flesh -from the beasts that will be roaming in the night.

She gestures toward the surrounding grasses in a way that tells Ar-gah' to gather bunches of it for bedding and warmth, then motions toward Bo-nah' and his need for restful sleep. Because of his wounds Bo-nah' will need help and support for walking. They also have to carry as much of their killed animal as possible. They will need to start their journey back to the sleeping place very early in the morning if they are to make it back there by nightfall.

=====

As she works her bone needle through the stiff hides, Yah'-nay sees her raven friend, Bw'-ak, circling above. She tosses a morsel of meat from her small stash of it. Bw'-ak settles alongside the morsel and pecks it up. The bird gazes at Yah'-nay, perhaps to see if there might be another morsel forthcoming. Seeing none, he flies away toward the hills to the west.

=====

Yah'-nay grows fearful as the day draws to a close, and there is no sign of the hunting party. She turns from her repair work on the hides, to grab some stones from a small pile nearby. She tosses them in the air and studies the pattern they form on the ground looking for signs and messages. This is unsatisfactory. She looks to the west, in the direction that Bw'-ak had flown. The bird didn't see the world in the same way as Yah'-nay, but his soft vocalizations in her presence were comforting to her. She tried her best to mimic each of the sounds as he made them. She hoped to understand whatever reaction Bw'-ak might have to her mimicry; perhaps someday to speak with him.

Once, when Yah'-nay and her sister were stripping the hide from a freshly killed small animal, Bw'-ak and several of his group landed nearby. They began vocalizing in diverse and complex ways. Of all the animals, Bw'-ak and all of the members of his group of ravens talked to each other in this way whenever and wherever they gathered. Yah'-nay felt a brief surge of longing to be with them to talk and share their stories.

Bw'-ak distinguished himself in his group, first, because he had a single white feather on the top of his right wing. Second, he dominated the behavior of the other ravens in his group with deep, rasping calls of anger. He began fluffing his head and leg feathers and flaring his ears whenever another raven approached a piece of food or pretty bauble that Bw'-ak had his eye on. The other ravens always deferred to Bw'-ak when he displayed in that way. Once his dominance had been established, however, Bw'-ak often backed away to allow others to have the food, or whatever other object had gained his attention.

Yah'-nay was pleased to see Bw'-ak's return to her side. They gazed at each other for a moment, and then she looked north in the direction the hunting party had gone. Bw'-ak flew off in the direction of her gaze. Yah'-nay turned back to the repair of the hides, but her mind wandered once again to the fate of the hunters. She put aside her repair work, and untied a small hide containing her personal belongings. She selected a necklace of small polished stones pierced and strung on a thin strip of hide. She hoped it would help her focus on the hunters to learn what progress they were making in their quest for fresh meat.

Yah'-nay sat down and faced in the direction of the hunters. She bowed her head and began a rhythmic hum while rocking forward and back on her haunches. She turned her head up and opened her mouth to call out to the skies as if asking for help in learning about the hunters.

The sounds woke Pe'-dee from a nap earned after finding some tasty roots. She looked at the old woman and grew fearful as her voice rose and her rocking grew more intense. She began to feel Yah'-nays intensity in her own small body. Soon Pe'-dah and Yah came to kneel beside Yah'-nay. They too began to chant, at first softly, then more loudly with the same kind of rising intensity. Pe'-dee began to cry as the three older women's chants grew stronger, and their rocking bodies began to shake with emotion.

After a few moments, as the sky darkened further toward night, Pe'-dee's cries turned to screams. This broke the spell that gripped the older women, and they fell to the ground, sweaty despite the cool air, and exhausted. Boa came up to comfort the women and cover their bodies with hides to warm them. Together they brought Pe'-dee into their arms to assure her that she was safe, and all was well with the group.

After eating the delicious roots that Pe'-dee had found earlier, the group gathered more tightly together and fell into a deep sleep.

====

In the gathering dusk, Ah'-oom followed Ar-gah' into the woods to inspect the fortifications in the area where they would try to give Bo-nah' some sleep time. She was concerned that animals would attack them in the dark, unless they could get off the ground and lodge some sharp sticks pointed downward, partly to protect the buried meat, and partly to protect them from attack by curious predators. The situation seemed impossible, but Ah'-oom felt she had no choice. She had to bring some meat back from the hunt, and she had to make sure that Bo-nah' could not only walk back to their sleeping place but would also have strength enough to help defend the hunters if trouble arose on the walk back. Bo-nah' was awake but still groggy from his wound. He assured Ah'-oom that he could do his share of the work. On re- gaining Ah'-oom's confidence, he turned back to the carving and shaving of the bone he would use to give himself a stronger and more accurate throw of his spear.

Bo-nah' had been working on this project for several moons. One day back then he watched some of the younger boys compete with each other to see who could throw a rock further. The size of the boy usually dictated the length of the throw. As usual Ar-gah' usually threw the farthest.

Then, one of the smallest boys picked up a rock and nestled it in a long piece of hide. The middle of the hide was widened slightly to more easily hold the rock. Carefully, the boy picked up both ends of the sling in one hand and began to twirl it over his head. When he felt he had enough speed, he released one end of the hide. The rock flew far beyond the farthest throw of the biggest boy in the group, Ar-gah'.

Bo-nah' took note of this. He wondered if the idea of increasing the distance from his shoulder to whatever projectile might be used would somehow give any projectile more speed and power. On another day, Bo-nah' began to wonder again about the value of giving the rock more speed and power. The rock remained an imperfect killing instrument. The killing or wounding power depended on accuracy in hitting the head of an animal in just the right place; difficult even for the best of hunters with the easiest of shots. It was a warm day, and Bo-nah' let his mind flow freely. He began to wonder if he could somehow use the piece of hide with a pointed spear. In hunting, a well-shaped spear with a sharp point had the advantage of being much more aimable and penetrating than a blunt rock could ever be. Bo-nah' tried to get his head around these thoughts, but the effort proved too much and his body gave in to fatigue and the warm sunny day. He fell asleep.

Later, he awoke and entered into a frenzy of activity. He was burning with a passion to see what would happen if he used the strip of hide to throw a spear. Clearly there was no way to twirl the spear with the hide as the boy had done with the rock. He tried an overhand throw and release with the spear nestled in the pocket of the hide, but, after several attempts, realized that he could never launch the spear with either power or accuracy using the hide.

Bo-nah', in deep frustration broke the spear over his knee, and threw the hide back at the boyish owner. The boy had been watching the entire futile exercise. Bo-nah' stomped away waving his arms and uttering a variety of growls and grunts.

Several weeks later, Bo-nah', once again, woke suddenly and with passion. The biggest problem was the strip of hide. He knew from experience that aim and accuracy in the throw of a spear demanded a straight, strong line from the heel of his back foot to the extent of his arm at the time of release. Whatever the distance between the releasing hand and the projectile – stone or spear - the penetrating power at the time of release depends on extending that straight strong connection through to the projectile itself. Bo-nah' would need to find a strong branch, or piece of bone to test this new theory.

He was sure he had found the solution, but it would take work to find the best piece of branch wood, to cut it to the proper length, to work out all the other details; such as the method for connecting the branch to the spear during the throw, and then release cleanly at the point of maximum thrust. Once the parts were assembled and shown to fit well through the throwing arc, and then require the clean release; it would require a lot of practice aiming the spear and throwing it for maximum effect. Bo-nah' had his work cut out for him.

Fortunately, Bo-nah' found a piece of leg bone in a stripped animal carcass that could be quickly shaped for the purpose of testing his ideas about the proper size, shape, and durability. One end of the bone looked like it could be shaped into the necessary connectable pocket to hold the spear. This was the bone that Bo-nah' took with him on the hunt with Ah'-oom and the others.

=====

Ah'-oom walked with Ar-gah' toward the area that he had fortified for the night ahead. Ar-gah' pointed toward the piles of wood he had gathered and piled around a tree big enough to support whatever they decided to build. Ah'-oom frowned at this pile, then looked at Ar-gah' and grunted her displeasure. She pointed to the sun sinking beneath the horizon as the evening skies darkened into night. She motioned him to come with her to plant the larger branches around the base of the tree, leaving enough room between the tree trunk and the branches for each of them to sit with some protection from the animals of the night.

Ah'-oom then pointed to the smaller sticks scattered around and directed Bo'-nay to gather enough wood for a fire that would burn through the night.

Once she and Ar-gah' had set some of the branches against the tree, she motioned Bo-nah' into the shelter to rest. Bo-nah' seemed to be recovering quickly, but he was glad that Ah'-oom did not give him any duties. He took the piece of leg bone into the shelter and began using a small rough stone to continue shaping the spear pocket at the end of the leg bone. The hook would hold the spear and leg bone together through the hurl and release. The hurl and release is the most difficult part of his idea, and he is desperate to test it.

Though each movement of his wounded arm – his throwing arm - caused intense pain and sweating, he was eager to test his idea, even if he would have to show one of the other hunters how to use it. Of the three hunters in the group, Bo-nah' rejected Ah'-oom because she would have other things to think about. He rejected Bo-nee' because his tracking skills would be needed to find the shortest way back. The third hunter, Ar-gah', was probably the best choice. His eagerness to prove himself on the hunt, would earn him the praise of the other hunters in front of the larger group. He particularly wanted Boa and Yah, his adoptive parents, to feel confident that he could repay the lifetime of comfort and support they had provided to him.

More importantly, Bo-nah' knew that the opportunity for taking training from Bo-nah' at such a critical time would give Ar-gah' a feeling of superiority over Ah'-oom. With the skill and more effective hunting weapon that Bo-nah' could give to him, Ah'-oom, because of her small size, could never throw farther than he could, given the same tools.

It was too dark by now to see his work, but Bo-nah's sensitive fingers could feel the imperfections in the pocket he is carving in the bone, and he used the stone to gouge and smooth further. The others had foraged as far into the woods as possible until it had become too dark. Ah'-oom motioned to Bo-nee' to build a fire and stack the wood nearby while she and Ar-gah' continued to stack the heavier branches around the tree. They were stacked thickest on the side of the tree away from the fire. A small opening toward the fire would make it easy to move from behind the protective branches to feed the fire when necessary.

Like all hunters in these days, Ah'-oom knew that none of the others would fall into a deep sleep while on a dangerous hunt. Bo-nah' was a possible exception, she thought, because of his wounds. Bo-nah' moved to the side away from the fire where the protective branches were thickest. Ar-gah' moved next him. Ah'-oom and Bo-nee' sat on either side of the opening the better to watch the fire, and to watch for any movements in the woods beyond. The two hunters fell into a light sleep.

Bo-nah' put the leg bone into Ar-gah's hand and then moved the hand over the spear pocket that had been working on. Ar-gah' knew immediately what it was. He took the bone from Bo-nah' and rubbed his hand over the full length of it. He took the bone in the hand of his throwing arm and waved it in the close confines of the protected area. It was as if he could not wait to try out this newest weapon.

Bo-nah' was pleased. He motioned to Ar-gah' to sleep, then motioned toward the sun's rising and touched Ar-gah's shoulder. Bo-nah', the effect of his wound catching up to him, fell into a deep sleep. Ar-gah', alert with excitement and anticipation, hoped that he would get at least some sleep before the sun rose once again on the wide Savannah.

He picked up the bone and a short branch that lay nearby. He ran his throwing hand over the bone and imagined the timing of the release of the spear at the point in the throwing arc where the spear would have its greatest accuracy and penetrating power. Ar-gah' knew that the most difficult of the tasks involved with the throwing would be the timing of the release. He continued imagining and practicing the throw and release of the spear without letting the bone drop for a long time.

=====

Night passed without incident. The hunters were able to sleep through the night, mostly undisturbed. As the grey light of morning began to lighten on the eastern horizon, Ah'-oom rose from her sleeping place and stumbled toward the now nearly dead fire. There were enough coals to get the fire going again and she did so, hoping to find some berries or leaves nearby suitable for making tea to help them on their way.

Looking beyond the file toward the path they would takeback to the main camp, she noticed a raven huddled against the cold on a tree branch a few paces away. The raven took note of her, and uttered some deep, rumbling vocalizations. Ah'-oom noticed that the raven had one white feather on his right wing. She recognized the bird as one of the many birds and animals who visited Yah'-nay most days. Ah'-oom took this to be a very good omen for their return trip.

She went to wake the others to begin assembling and organizing the weapons, the meat, and food supplies they would need for their journey. They needed to get underway.

Bo-nee' separated two of the longest branches out of the woodpile. These would be used to carry supplies, and, if necessary, Bo-nah' if he became so tired that he could not continue walking on his own. When Bo-nah' came out from among the protective branches around the tree, however, it looked as though he had completely recovered. He gestured to Ah'-oom and Bo-nee' that he was ready for the trail ahead, and they should get started.

Ar-gah' appeared, smiling, and carrying the leg bone that Bo-nah' had been working on for several weeks. Bo-nah' picked up his spear and handed it to Ar-gah'. Ar-gah' turned away from the group and pointed at a tree about 20 paces away. He held the legbone in such a way as to fit the back end of the spear to the end of the bone opposite the end he held with his throwing hand, then with his free hand he placed the spear along the top of the bone, then grasped it with the same hand he used to hold the bone. He lifted the spear and bone and cocked his throwing arm back as though preparing to throw.

With one quick move, Ar-gah' threw the spear. It missed the tree he aimed at by a little, but it completely pierced a smaller, further tree a little to the right of the target. Ar-gah' turned back to the hunters with a big smile across his face. Bo-nee' and Bo-nah' whooped and shrieked with pleasure. Bo-nee' ran to Ar-gah' and threw him to the ground in celebration. As they rolled around, Bo-nah' wanted badly to join them, but held himself back rather than risk further harm to his throwing arm.

Ah'-oom took note of the success of this demonstration. She expressed her great pleasure at the addition of the new, and very effective weapon to their small arsenal. She jumped into the rolling pile of bodies and gave as good as she got in return when the punching, kicking, and screaming began in earnest.

Eventually, out of breath, she got up and rested. Then she shouted and gestured that the time had come to go. The four hunters gathered their gear and loaded the pallet prepared by Bo-nee'. Bon-nee' assumed that Ah'-oom would direct that Ar-gah' and he would carry the pallet, but she said she would take one end.

She gestured that Ar-gah' should go ahead of them, and practice walking, and running, and throwing the spear with the leg bone. She showed how he should always keep the group in sight. She pointed toward the sun and indicated to Ar-gah' that he should stop and wait for them when the sun had moved about 30 degrees; a distance she demonstrated by pointing along the sun's forward path with her throwing hand. She could foresee many possibilities for violence on the trail ahead that might require the use of the new weapon, and she wanted Ar-gah', now the strongest thrower among them, to be ready for all possibilities.

She set a pace that would be best for Bo-nah'. Bo-nah' protested that he felt fully recovered, and she should not slow the others down when he was sure he could keep up. She walked to him, put her hand on his arm, and looked into his eyes. He recognized her determined look and looked away. Ah'-oom returned to the head of the pallet, picked it up in unison with Bo-nee' and motioned for the others to lead the way. With Ar-gah' and Bo-nah' ahead of her, she would be able to assess Ar-gah's progress with the new weapon, and she could watch Bo-nah' for any signs of fatigue.

They made good progress as the sun began to break the horizon. Yesterday was a surprisingly good weather day, and today was looking good as well. Two days in a row at this rainy time of year would be a needed and pleasant surprise.

Ah'-oom looked up to check the clouds and weather. She saw the raven above them. She could not see if it was the raven with the white feather, but she continued to believe that the raven's presence with the travelers to be a very good omen for the success of their return to the main camp. She pointed to the raven to make Bo-nee' aware of it. She gestured and spoke Yah'-nays name indicating that the bird was a friend of Yah'-nay and was watching out for them.

=====

Yah'-nay awoke to the new day after a troubled night. She felt fearful about the fate of the hunters, but hopeful about their safe return. Pe'dee had been looking for roots, grubs, and other tasty morsels for those in camp to share. She approached Yah'-nay with hands full of these succulents, and with a selection of pebbles that could be polished and used to reward the hunters when they returned. She then went on to build a fire for a tea she would prepare from the leaves she had gathered.

Boa approached to take some of the food, then indicated that he would go to a lookout point further up into the hills. From there he would have a better view of the valley, and the trail that the hunters would follow on their return.

Yah'-nay, Pe'-dee, and Pe'-dah would remain behind to find more food to satisfy the hungry hunters on their return. Yah'-nay also knew that it would be important to have a valuable gift and songs ready to reward the most successful hunter. She knew that making such a gift would be her job and she was grateful that Pe'-dee had brought such pretty stones for her to work with.

Boa assured the women that he would also keep an eye on the camp should any trouble arise. If there was trouble, he would not be too far away. Yah'-nay wondered about the flock of ravens that she had watched flying away along the trail of the hunters. The raven with the white feather had also flown in that direction, but none had returned. Yah'-nay knew that her patience and calm would be important to the others through this troubled day, as all awaited the return of the hunters. She settled back and picked up a stone to polish. She would polish by rubbing off the dirt with another stone, then rinsing with water from the creek.

=====

By midday, when their body shadows were shortest, Ah'-oom knew she and the others would have to stop. Carrying the pallet was hard work, and the weight was cutting into her shoulders. Besides, it was time. The sun had reached the point in the sky that she had indicated to Ar-gah' and he was already walking back toward them with the legbone and practice spear.

As Ar-gah' approached the others he seemed very pleased with himself. By her hand gestures, Ah'-oom indicated that Ar-gah' should show them his progress in acquiring this new-found skill with Bo-nah's weapon. Ar-gah' stood and pointed at a bush at least 50 paces away. He settled the spear into the pocket Bo-nah' had carved in the leg bone, then cocked his arm ready to throw. With what seemed more like a flick of his wrist, the spear flew higher and faster than any of the other hunters, even Bo-nah', had ever seen. It flew directly to the bush, pierced the middle of it and planted itself deep into the ground behind.

A bit of dust stirred up, but the watching hunters thought it might a puff of wind. Ar-gah' walked to retrieve his spear and picked up a rabbit that the spear had penetrated. He did not say whether he had known the rabbit was behind the bush before he selected his target. Though too tired to join each other in a screaming, shouting wrestling match, the three waiting hunters uttered a series of sharp barks and howls to indicate their great pleasure at the progress Ar-gah' had made and particularly in the surprise and welcome gift of fresh meat.

The hunters stripped the skin off the rabbit and ate the meat raw. There was no time for the pleasantries of cooking. Then they covered themselves from the sun with hides so they could take short naps. Bo-nah', once again, offered to take Ah'-oom's place carrying the pallet. She waved him away until she had napped. Before covering her head she noted that the raven that had been following them had flown away to the west, and had not returned. She was concerned about the absence of the raven.

Some time passed, and Ah'-oom suddenly jerked awake. She went to the other hunters to kick them awake. She barked at them to make them wake up and get moving. She did not want to lose more daylight and be forced to spend another night away from their main camp. All were groggy, but they assembled their gear and began to move. Bo-nah' blocked Ah'-oom to prevent her from picking up her end of the pallet. She did not complain.

She noted that Bo-nah' seemed to have fully recovered, but he did give out a yelp of pain when his wounded arm bore the weight of picking up his end of the pallet. Nevertheless, he motioned her ahead to act as Ar-gah's eyes and ears along the trail. In addition to B-nah's leg bone, Ar-gah' had three of Bo-nah's best spears, leaving four for the other hunters. Ah'-oom still carried Boa's ax and her own spear and short, bone, cutting tool for defense. If they found some spear-size branches along the way, they would stop to make at least a few more spears.

They were coming into some broken country where water sometimes flowed after heavy rains. Judging by the dark clouds on the horizon where the sun normally rises, the possibility of flowing water, perhaps a flood of it, became very real. There would surely be animals and there might be other hunting groups to be dealt with. They kept moving to the south with deliberate speed. Bo-nah' was as good as his word and did not slow the group down. He did not complain about the weight of the pallet, even though it was clearly causing him pain in his wounded arm.

Ar-gah' was moving quickly, and it was all Ah'-oom could do to keep up. After a time she had to call a halt to wait for the other hunters. Ar-gah' indicated he would find a high place and keep looking around the area. He would return shortly. Ah'-oom was grateful for his energy and enthusiasm in watching out for all of them.

The two hunters with the pallet joined her and paused with her as she turned her body to look toward all the horizons ahead of them; to sniff and taste the air; and to shush the hunters so she could listen for noises that seemed out of place. She did hear the call of a distant raven but did not think it important enough to call attention to it.

=====

Yah'-nay looked toward the southern hills upstream from their summer campsite. The clouds on the horizon there were growing dark and ugly. She went to Boa on the lookout point and called his attention to the possibility of heavy rain, and the possible flooding of the creek. Boa looked troubled, but he quickly resolved on a course of action. He motioned to Yah'-nay to get the others out looking for shelter materials and dry firewood.

A stormy night would cause the sky to be very dark. They would have to build a shelter structure from small trees covered with hides to keep the rain out. They would have to build a fire in the shelter large enough to guide the hunters back to the camp where they would have shelter and food. They would be very tired, and very hungry from their long trip. Yah'-nay knew that she would have to make sure that they had comfortable places to sleep near the fire.

=====

Ah'-oom suddenly stiffened. She turned her head and body to the west. Her eyes, ears, and nose focused on something in the distance. She gestured for absolute silence. She pointed. The two litter carriers strained to see what she saw. Then, in the far distance, they saw some shapes. Ah'-oom needed to figure out if they were moving, and what direction.

After some time she looked at the other two hunters with a look of concern and questioning. She gestured in a way that asked which direction the shapes were moving. Bo-nah' and Bo-nee' both pointed to the south… the same direction they were traveling. Their paths would eventually cross. Ah'-oom gestured that the two hunters should stay very low, but alert while she and Ar-gah' went to investigate the other travelers. She picked up one of their spears and balanced it in her throwing hand. She indicated the other spears and indicated that they should all be checked by the two hunters.

Ah'-oom moved toward Ar-gah'. He had seen the shapes and noted their movements. He was crouched and watchful. Ah'-oom crouched quietly beside Ar-gah'. Once again she moved her head and body in a way to better get a feel for the atmosphere, the weather, and the movements of the shaped figures.

The figures were still moving toward the south. If they continued in that direction they would eventually cross paths. Ah'-oom consider the possibilities of interaction with the moving shapes. They might be peaceful hunters like themselves, or they might be thieves looking for groups like hers with weapons and meat that could be taken away. Either way, if her group did not interact with them in some way, that group might be able to get to the sleeping place before Ah'-oom and her group. There were more of them than of Ah'-oom's group, and they appeared to be moving faster across the ground. She pondered further. If the strangers were hostile Boa, Yah'-nah and the others might need all the help they can get.

Ah'-oom heard a soft warbling call from the raven overhead. She looked up and saw the raven appear to look back at her. The bird then turned to the west toward the strangers. She stood up and put her hand on Ar-gah's shoulder. She looked in his eyes and indicated that the two of them would go out to meet with the strangers.

Ar-gah' frowned at this. He stayed crouched for a moment, but Ah'-oom slapped him hard on the head. She started to move toward the strangers. Ar-gah' growled under his breath and barked at her. He paused to consider what Boa and the others in the sleeping place might do to him if the hunters returned and had to report that both Bo-nah' and Ah'-oom had been injured on the hunt, and he had not been hurt.

The killed animal they were bringing back might not look like much of a reward for their efforts. There would be no songs of praise.

He then thought that Ah'-oom might be seriously wounded or killed by the strangers. As it crossed Ar-gah's mind, his body shook and he rubbed his forehead vigorously to purge the thought. He got up and ran to catch up with Ah'-oom. On his approach, she motioned him to stay a few paces behind. She pointed to the leg bone and spear, then indicated that he should hold the spear as a hunter would and keep the bone out of sight. Ah'-oom knew that she and Ar-gah' would be outnumbered by the strangers. She hoped they would be seen as peaceful hunters rather than threats.

By the time they were within calling distance, the sun had moved further in its arc toward the horizon and night, and the dark clouds to the south had continued to grow toward them.

The strangers now stopped and looked toward the two approaching hunters. Ar-gah' continued to hold his spear as a hunter would while looking for game. Ah'-oom had taken the leg bone from him and put it in the strip of hide around her waist next to the ax. This way, she hoped it would look less like a weapon. Ar-gah' knew that he might have to quickly take the bone, fit it to his spear, and throw it with precision if their interaction with the strangers began to go bad.

Ah'-oom walked to within a few paces of the strangers and stopped. She gave two soft barks of greeting in their direction. She noted that there were less of them than the fingers on her two hands. All carried spears, but they did not seem to be threatening. Two of the strangers carried a piece of hide wrapped around several long pieces of wood. The two males also carried a long, thin piece of wood with a thin strip of hide tied to both ends forming a kind of bow. She did not know what to make of this.

There were three women in the group. Two carried branches used to build shelter when necessary. The third woman carried a hide slung over her shoulder that appeared to be used to carry food stuffs. One thing about the group of strangers was very peculiar to Ah'-oom. All of them except the oldest had very little hair on their bodies.

Ah'-oom's group were mostly hairy. She was the least so, and it caused her to feel the cold air much more intently than the others, to the extent that she needed extra hides at night to be able to sleep. Also, when the strangers spoke to each other, they spoke in ways that Ah'-oom felt she could not shape her mouth to speak, and the others in her group were even less able to make the sounds the strangers were making. The strangers spoke among themselves in very measured, rhythmic ways; ...almost like they were singing to each other. Ah'-oom became fascinated with their speaking. She stepped forward to more closely watch the way their mouths worked as these beautiful sounds came out.

She suddenly stopped and looked up in time to see the two males with the thin strips of wood bowed by thin hides stretching the bow and hides to aim the sharpened long sticks at her. Ar-gah' was immediately at her side, she stopped him before he could reach for the leg bone. She motioned him to stay behind her. Then she put both hands, palms out, in the air, and bowed her head to the ground in an act of submission. Then she kneeled, still with her hands in the air, and her head bowed. Ar-gah' understood the need to show submission, but he only bowed his head down, and opened his palms toward the strangers.

The oldest and hairiest of the strangers then approached them both. He put his hands on both their upper arms. He invited Ah'-oom to rise. She did so, and both looked into the eyes of the old man. The stranger motioned to one of the women in his party carrying food. She came over and took her sling of food off her shoulder to hand to him. He picked out some succulents and offered them to Ar-gah' and Ah'-oom. The two hunters accepted them and ate them.

The old man then bade them all to sit and share their meager store of foods. Ah'-oom tried to convey the existence of Bo-nah' and Bo-nee', and her need to rejoin them. Eventually, together, she and the old man agreed that Ar-gah' would rejoin the other two hunters, tell them what was going on, then help them get on the way back to the sleeping place. Ah'-oom would stay with the travelers. As a sign of mutual respect two of the men with the traveler's party would go with Ar-gah'. They all knew that they must then try to get to the sleeping place where there would be shelter and fire against the coming storm.

With gestures Ah'-oom urged the strangers to prepare quickly for fast travel toward the hills in the south. She would lead them to the sleeping place. Once she had picked up her own stuff, she picked up one of the bundles of hides the others had brought to fend off the cold. As she was with the hunters at the beginning of the hunt, Ah'-oom moved to the fastest pace she could find that each of the strangers could match for the time it would take to return to her family at the sleeping place.

As the darkening clouds in the south grew more threatening, Ah'-oom heard the call of raven as the bird flew low over the group, appeared to glance in her direction, then continued south.

They were on their way.

====

Boa pulled the hide tighter around his body as a chilly wind had been building up ahead of the storm. He had been looking north and hoped he would not miss seeing the returning hunters as soon as they came within his vision.

Yah'-nay joined Boa. She put another hide around Boa's shoulders, and gestured to the new, partially buried shelter built by Yah'-nay, and Pe'-dah.

Boa was pleased to see the other piece of carefully tanned hide that Yah'-nay presented to him. It is a piece of bright blue stone held in place on a field of ocher-colored hide by thin pieces of hide blackened by smoke and careful burning.

Ah'-oom's young sister Pe'-dee sat beside the elder healer. She leaned her head into the healer's breast. Yah'-nay put her hand on Pe'-dee's head and on the blue stone to indicate to Boa that the child had found the stone.

Despite his pleasure Boa was growing more concerned as time passed and daylight faded. He indicated to Yah'-nay that the fire needed to be started. He hoped there would be enough wood and fire material to last through the night if that became necessary. Yah'-nay assured him that the fire will be built in time, and there are plenty of materials to hold the fire through the night.

As they sat quietly together the raven with the single white feather on top of its right wing came to them and landed a few paces away. The raven hopped sideways a little way toward Yah'-nay then vocalized various sounds that Yah'-nay recognized as sounds of comfort. She smiled to herself and chanted softly in a way that would give assurance to Boa that all preparations were as good as they could be.

Eventually, the fire was lit while there was still an evening glow from the setting sun. Boa could no longer see very far in the direction of the hunter's return. He hoped they could see the fire. Shortly after, Ar-gah' appeared followed by the other four hunters bearing the palette with their weapons, and the animal they had killed the day before. As soon as the hunters came in sight, the women began to sing a barking ululation to indicate their great pleasure at their return.

The younger ones ran down the hill to greet them. Boa, Yah, and Yah'-nay descended more carefully being mindful of the hazards on the darkened path. Boa grew immediately concerned that Ah'-oom was not with them. When Ar-gah' realized that she was not among the greeters, he too became concerned. He and Boa both turned back to the trail to begin a search for the missing. Boa put his hands on both hunters when they approached to caution them against going back out in the dark, stormy night. He indicated that a better plan would be to pile more wood on the fire to give the missing a bigger target in the dark.

Ar-gah' looked at the gathering storm and wind. He agreed with the wisdom of Boa. The possibility of the two groups missing each other, and becoming more lost, was very great. The group moved back up the hill to gather more wood for the fire. The women moved into the shelter and began a soft chant to give guidance to the travelers.

The fire spread warmth throughout the shelter. The men also gathered around to discuss the results of the hunt.

Boa took note of Bo-nah's wounded arm, and the hide Ah'-oom had applied to it. He pointed at the wound with a questioning look. Bo-nah' related the story by gesture, and by showing Boa the site of various blows and wounds inflicted on the animal that now lay on the nearby palette. He sought permission to lay down to rest as he had pushed himself very hard to help bring the animal and their weapons back to the group's sleeping place. He especially praised Ah'-oom's leadership on the hunt.

Ar-gah' showed Boa the leg bone that Bo-nah' had prepared. He picked up a spear from the palette and fitted it to the carved out niche. He demonstrated how to hold the two pieces but did not throw the spear. Bo-nee' had sat down, but he was able to express his great pleasure at the skill that Ar-gah' had learned in mastering the use of the bone to throw the hunting spears with more power and greater accuracy.

Bo-nah' then lay back and appeared to fall asleep. Yah covered him with a hide. She touched his forehead and his wounded arm to decide if she needed to do more.

As Boa looked on with concern, she assured him that Bo-nah' would be fine with sleep. Ar-gah' indicated the two travelers who had remained at the edge of the firelight. He described the experience he and Ah'-oom had in their meeting with the strangers. He told them both that Ah'-oom had stayed with them and would be leading them to the sleeping place. Boa looked concerned, but Ar-gah' indicated that he and Ah'-oom had shared food with them. He believed they were travelers who had their own food and shelter and would cause no harm. He described their absence of body hair – more like Ah'-oom – less like he, Boa, and other members of their hunting group. He also described their peculiar way of talking; …more like singing than the barks and guttural vocalizations of their own group.

Now, Ar-gah' motioned them forward. Each lay his spear down, then approached Boa and extended a hand in greeting both touching Boa's upper right arm. Boa responded with a low grunt. He bade the two strangers sit with him. He motioned toward Yah and Yah'—nay that they be fed.

Rain started to fall. The roof of the shelter started to leak, and the men began moving hides around in the hopes of keeping the inside of the shelter as dry as possible. The fire in front of the shelter was built up strong enough, it was hoped, to persist even though a heavy rain. The women continued their chant in the hopes that the returning travelers would hear their call.

Bw'-ak the raven flew from his perch on the shelter roof toward the north in the direction of the returning group. Yah'-nay continued her soft chanting as she watched the bird go.

Ar-gah' said he would keep the fires going. After a while most of the members of the group fell into a deep sleep.

Later, while scanning the dark beyond the firelight, Ar-gah' spotted Ah'-oom walking alone along a path to the sleeping place. He got up and moved quickly to her, being careful to avoid the broken ground that could cause serious injury. Ah'-oom indicated that the others were coming, but their burdens had grown heavy. The two of them went back to the travelers to help. Once they had re-distributed the loads the two hunters and the travelers made their way to the other hunters, and to the fire and shelter.

Ar-gah' went to Boa and Bo-nee' to tell them of Ah'-oom's return with the travelers. Ah'-oom woke Yah, Yah'-nay, Pe'-dee and Pe'-dah. Once awake the women once again began the ululating barks of joy at the safe return of all the hunters. Boa and Bo-nee' hugged the returning hunters and bade them sit down by the fire.

The women who had been chanting were quiet now. Yah'-nay approached Boa and gave him the beautiful blue stone with the colored hide backing and weaving that she had shown him earlier. Boa was momentarily surprised. Then, he called Ar-gah' and Ah'-oom to his side. He looked into the eyes of both of them for a few moments, then he put the decoration on the chest of Ah'-oom. She was overwhelmed. She bowed her head then hugged her father. She turned to look at Ar-gah'. He had no expression in his downcast eyes.

She stepped toward him. Then she put the honored decoration she had received from Boa, her father, on the chest of Ar-gah'. His expression changed to one of joy and pride that he tried very hard to suppress.

Bo-nah', who had woken up, looked at Boa and bowed his head toward his father to indicate his approval of all that had occurred. He looked at the travelers now joined with his family around the fire. He knew they were from the north, but he wondered where in the north. He began to imagine places filled with many wonders. He resolved to go there some day, perhaps with the travelers when they returned north.

The sky along the eastern horizon had begun to turn gray with the emerging sun.

PANEL 2 - COMMERCE

Memphis is a trading center on the Nile River between the southern valley and northern delta where it crosses the overland route that connects western North Africa and Mesopotamia. Hesina, daughter of a successful commercial warehouseman, awaits a caravan arriving from an oasis in the western desert. The caravan carries leather goods, dates, fruits, and other trade goods.

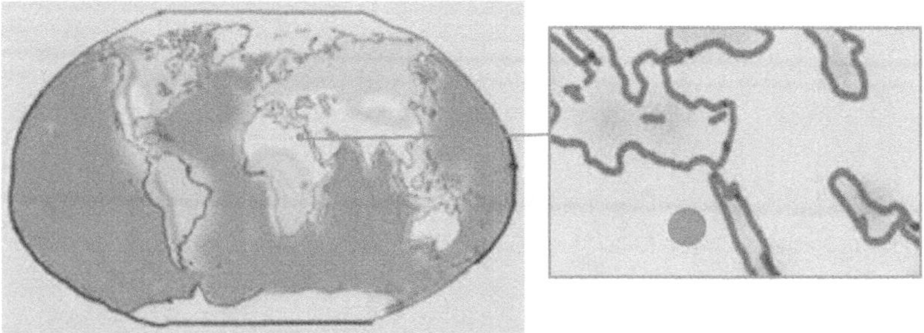

MEMPHIS, NORTH AFRICA. LATE SUMMER, 3,000 YEARS BCE

PROTAGONIST

Hesina – A youthful girl talented with numbers, with new ways to count, with the recording of transactions, and with the schmoozing with traders who do business with her father's trading and warehousing businesses.

Senna – The Father of Hesina who supports his family as a builder and owner of warehouses for the temporary storage of the huge quantities of trade goods that pass through the area.

Horus – The name Horus is more fully given as Horus the fighter. Horus is an Egyptian diety known as the God of the Sky, wind, and kingship.

Ptah – The Egyptian God of craftsmen and architects.

ANTAGONISTS

Alim – An older trader on the caravan routes between Africa and Mesopotamia who has pledged to out-bargain Hesina with his load of highly prized dates;

Rabiah – A son of the older trader who has designs on Hesina.

LESSER CHARACTERS

Salama – Mother of Hesina; wife of Senna.

Nena – A friend of Hesina, and consort of the Pharoah.

Serenen – Hesina's personal slave, protector, and mentor. Castrated because he brought offense to a former master when he was caught in an embrace with one of the Master's concubines..

Heti – An upriver farmer, old friend, and business associate of Senna and Hesina.

Akar – A boatman on the Nile. He and his crew of rowers and sail-setters take Hesina upriver to Beni Hasan.

Paru – A palace guard assigned by Pharaoh to accompany Hesina and her retinue upriver to Beni Hasan.

MINOR CHARACTERS

Iaret – Senna agent in Jerusalem.

Faiza – A jeweler and trader in Lapis Lazuli from Ur, a large city in Mesopotamia.

The City of Memphis lays at the south end of the braided outflow waters of the Nile Delta. The Mediterranean Coast lies 120 miles to the north. The north border of the area known to the Egyptians as Nubia lies 650 miles to the south. Far to the east across the ancient and rich agricultural arc known to us now as the Fertile Crescent lies the Crescent's easternmost large cities, Babylon, Uruk, and Ur along the Euphrates river near the mouth where it empties into the what we know as the Persian Gulf.

Equally rich and productive in agriculture are the farming areas of the delta, and on both shores and sloughs along the Nile. The main transportation corridors for moving the produce of the Crescent lays along the Euphrates river valley, then down near the east coast of the Mediterranean Sea. The main transportation corridor for moving the produce up and down the Nile river is the Nile river itself. Both transportation corridors meet at the city of Memphis.

Growers and their representative traders of agricultural products and other kinds of produce from these two productive regions needed warehousing and all the services required to secure, store, and move products from one owner to another.

As trading activities increased, commercial services designed to keep track of commercial transactions among the many traders and their many kinds of products bought and sold in many quantities became a matter of urgency. This kind of urgency drives innovation. This story is about some of that innovation in the City of Memphis, in Egypt in 3000 BCE before the unification of the Upper and Lower Nile valley and the beginnings of the first dynasties of the Pharaohs.

It is late summer. The Upper Nile River has returned to its banks following the annual spring flood.

Hesina, bright and pretty daughter of the prominent warehouseman and trader, Senna, sits cross-legged sipping hot tea on the roof of one of her father's warehouse in the western suburbs of the city of Memphis. It is dawn in Egypt. The early morning light from the east casts a weak shadow of her body as it falls across the roof in front of her. She stares intently to the west as she waits for Alim's caravan. He will bring baskets loaded with dates, nuts, and precious, but perishable figs from an oasis two days journey to the west.

She had been alerted to Alim's arrival by his son, Rabiah, who serves his father by acting as runner and messenger in advance of the arrival of his father's caravan of burros, dromedary camels, and other beasts of burden, all carrying items for trade.

Hesina knows that if she and her father can do trades with Alim for the figs, they will have to distribute them to their vendors and most favored customers in Memphis before they spoil. Among his other, less perishable products Alim may be trying to trade are dates. If Senna decides to buy these less perishable products, he and Hesina will have to find a company to ship the dates north along the Nile through the Delta to the Mediterranean port of Rhakotis, and from there by sea north to the ancient City of Jerusalem. There, the dates will be accepted and passed along to other traders and merchants by Senna's long-time agent and business partner, Laret, a Greek born and raised there.

Rabiah's father, Alim, is a tough trade negotiator, but he was not happy the last time he left a trade negotiation team with Senna, and, now his daughter Hesina. Senna had early recognized his daughter's talent with numbers and hieroglyphs, and with the hieratic script used in Egypt among the scribes to the highest classes. He had enhanced her natural talents by having her trained in being a scribe by the priests in the temple of Ptah.

Several of her instructors had told Senna of his daughter's many achievements at the school. Since then he has relied on Hesina's talent and training in languages and writing in all his negotiations with traders. Her skill with the new ways to tally sums using pebbles and grooves in the ground, had gotten Alim to agree to compensation that Senna knew was not much above his actual costs of delivering figs on his next caravan.

Alim did not realize this until he was a few days into his journey back to his family's encampment near the Fayum oasis in the western desert.

At the time, Rabiah had asked Alim, in Arabic: "Are you happy with the business arrangement with Senna?"

"No. The trade goods we earn are barely enough to justify the time and expense of organizing the next delivery of figs. It is worth the trouble, however, because we need to continue the business arrangement. We will think about how to get the better of him, and his precocious daughter with her pebbles in dirt the next time we do business."

Hesina wondered if she should bring her father, Senna, into any conversations she might have with Rabiah. Rabiah, though a rough-hewn farmer and trader like his father, is a very good-looking man. Hesina – only a few years beyond her bleeding - knows that she might be swayed by his rough charms if she spends too much time alone with him. On this morning she wore a shawl to fend off the morning chill, but most Egyptian women took off most of their clothes, except for a thin cotton wrap around their waist, when outside during the heat of the day. Hesina was still too shy to do that when away from her family.

Hesina's mother, Salama, did not try to alleviate this shyness. Every day, she reminded Hesina of her beauty, and how the bad men in the city would probably grab her breasts and other parts of her body if she exposed herself in public.

Hesina knows that her technical skills are very good, and she is a very good scribe. She had provided the tools of communication that had enabled her father and herself to get the better of several other traders. But the nuances between Senna and Alim were at least as valuable, and therefore at least as important to any trade negotiation because of their many years of doing business with each other. The new ways of counting and recording items of exchange were a challenge each time because there was so much detail involved, and the methods of counting and recording were constantly changing.

Hesina considered herself a master of these methods and her ability to teach her father and her personal slave and protector, Serenen, how to use them had made them, together, into very effective traders.

Quantities exchanged among several traders remained a mystery, however, but Hesina had resolved to try to find a better method of keeping track of items traded in quantity by a particular trader. She knew that such a method is essential in order to assign the profit of any trade to the correct trader.

In order to judge the value of a trade, items on both sides of each trade had to be identified as to weight, appearance, overall quality, and value. Unfortunately, there was no reliable way to store exchange information from one trading event to another, or from one trading season to another. Without stored information each trade was new, and each pattern of trading by individual traders is difficult to remember from year to year.

Papyrus reeds were plentiful along the Nile, and the production of scrolls of papyrus for record keeping and export was a profitable industry in most Nile communities. Papyrus, though the reeds were plentiful, and easy to make for record-keeping, use, and transport was somewhat fragile when handled by many people. Hesina believed it necessary to find a more permanent and more sturdy way to record some trades where conflict was likely in the future. One new method from Samaria in Mesopotamia involved the use of wet clay as a medium for a system of trading codes that could be punched into the wet clay using a triangle shaped stylus carved from wood or bone .

Hesina thought she might be able to design a specialized stylus with the same hieroglyph she used on papyrus to mark each trade by a particular trader, but she had not had time to pursue the idea.

Keeping a record of trades by making impressions in wet clay had been in use for some time in the distant cities along the Euphrates River in Mesopotamia. Hesina had learned of the clay tablets from a trader who had brought jewelry from the city of Uruk far to the east. The trader had come to Memphis to trade his jewelry for a quantity of cotton grown along the Nile; the kind used to produce cloth by weavers in Memphis.

Memphis is very productive in cotton because of the very fertile overlay of soil laid down each year by the annual Nile flood. Once the jewelry trader was able to complete a deal with Hesina's father for a large quantity of cotton, she had watched as the trader pressed his codes into the soft clay of the tablet. Once dried in the sun, the trade information could be stored permanently; even longer than those recorded on papyrus. As he bent to the tablet his coat fell open to display a neck piece with a hanging pendant made of an inset jewel of deep blue streaked with gold. In the eyes of Hesina the stone was a beautiful piece of jewelry

Hesina's attention returned to the clay tablets. She could see the value in having these impressions in clay if she could learn what each of them meant. She knew she had to work hard to learn the coding involved in using the information in the clay impressions.

In addition to being a very pretty young girl, she was also a very socially engaging one. She was able to persuade the Uruk trader to show her the language for the codes he used to record his trades. Hesina understood that the difficulty in using the codes lay in the lack of a standard language for the creation of codes. Without a commonly understood language for making standard codes the clay impressions could not be made unique by each trader for their personal and private use.

Hesina understood that the solution to this problem meant that each Egyptian hieroglyph would have to be associated with the same tablet symbol. A very daunting task.

After all, if a trader could not secure the information in his trade records, and keep them from being read by others, how could he trade in a way that gave him the upper hand over the person he trades with.

She thought about this for a few moments, then turned away from the trader. She remembered the bright blue and gold pendant he wore around his neck and resolved to ask her father about it.

=====

Hesina's extensive experience in recording using the hieratic script that combines the more formal hieroglyphics with an easier to use symbology worked well enough on papyrus, but counting items, and estimating the value of a transaction involving many items could not be easily done with hieroglyphics. Hesina hoped that the use of coded impressions in wet clay would help in these more complex computations once she learned how to use the codes.

She had also begun counting using a series of 5 to 10 shallow trenches on the ground, each parallel with the other, each about 2 hands in length. She had also selected a set of pebbles for the same size, weight, and color. She counted in multiples of 10 because she had 10 fingers, and the number system in hieroglyphics and hieratic had symbols based on 10 that are used to count items and record sums.

Use of the trenches in counting starts with placing an item In the rightmost trench which would represent the counting unit. An "item" could be a bundle of papyrus reed, or a basket of grain, or a hand-full of dates. Once she determined the individual item, she could count up to 10, or 100, or 1000 items as one unit for the units trench on the far right depending on the quantities involved.

In the leftmost trench she would place a pebble indicating that 10 units had been placed in the rightmost trench. In counting beyond 10 units, she would take a pebble from the rightmost trench and place it in the adjacent trench to the left. On placing the tenth pebble, she would add another pebble to the leftmost trench. This process would be repeated until all items had been counted.

Once the counting is complete, the appropriate numeric hieroglyphs and hieratics are marked on a sheet of papyrus, and the next counting task can be begun.

=====

Her father, Senna, encouraged Hesina in all her explorations and experiments with these new ways of recording values, and doing trades. He also enjoyed exploring new ideas for his several trading groups with her, and he hoped Hesina would continue her enthusiasm for learning ways of record keeping suited to whatever trading venture he might choose to get involved in.

With the recording systems worked out with his daughter, Senna had already built a substantial trading and warehouse activity by working with traders and shippers who needed storage space while small groups of them discussed their trade goods. In this way they could set deben values – a value set by a community of traders on something thought to have a commonly accepted value in trading - on their trade goods .

With this common value set as a deben, they could trade advantageously with traders bringing goods from other areas. These could include trade goods from the river's delta to the north for grains brought down the Nile from the south. Trade goods also came by caravan from the oases in the western deserts.

Some animal traders with their herds also came from as far away as the city of Uruk in Mesopotamia; 4 to 6 months of overland travel distant from Memphis.

His warehousing business was well-enough established that Senna had begun to explore a new venture in boat-building and shipping. He had set up a workshop In one part of the complex of buildings he owned in a western suburb of Memphis, and had engaged the services of a man who had designed some of the successful boats capable of carrying grain and livestock to Memphis from the irrigated agricultural areas along the upper Nile. Because he had already established warehouses for grain in his complex, he would not have to pay for storage while he negotiated with grain buyers if he owned his own boat.

Even better, he would be able to run a boat continuously during the harvest season in the late fall and winter because he could unload the grain for storage on his terms. He would not have to be under pressure to negotiate for storage space while his boat sat in the river under expensive security because his load of valuable grain could not be immediately landed.

Senna had another goal in mind for the new boat. He knew that the shipping route between Memphis and the northern saltwater port at Rhakotis was even more lucrative for a smart shipper and warehouseman like himself than was the route between Memphis and the grain and cotton growing areas upriver. Regular shipping among the many braided and irregular waterways of the Nile Delta would require far more care in boat design and building than would be necessary for a boat confined to southern trips on the Upper Nile.

The annual flood of the Nile in the summer produced so much water running so fast that well-established waterways through the Delta between Memphis and Mediterranean salt water in the previous year could change to channels that were too shallow for any commercial shipping after an annual flood event. Senna began to realize that he needed a much greater understanding of navigation among the Delta waterways, as well as more detailed knowledge of boat design. He wondered if it was even possible to build a single boat to travel both routes.

Senna wanted to talk with his daughter Hesina as soon as possible concerning these boat plans. Earlier that day, she had told him of the success of the trade of jewels for cotton with the Mesopotamian jeweler. She had also asked to talk about the beautiful blue and gold pendant worn by the trader. Senna felt there might be an opportunity with the Jeweler and wanted to talk to her about his plan for acquiring more of the raw ore for the valuable blue stone. When the ore is processed into the jewelry desired by the very wealthy families of Memphis, they can be persuaded to trade many valuable items for it.

====

The city of Memphis, as a cross-road for the major trade routes between the North African desert, the fertile Nile River basin and delta, and the great cities of Mesopotamia, continued to grow into a major commercial city. The City had a long history, but growth accelerated when Narmer – the great man who merged upper Egypt along the southern Nile, with lower Egypt along the northern Nile and the very productive Nile Delta to the north – proclaimed Memphis as his capital. He built his temple to the God of all craftsmen, Ptah, in that sacred place.

Commerce and enterprise began to accelerate after the merge. The entrepreneurial spirit among local traders – already very diverse and robust - quickly began to expand and traders began to gather to themselves all the quantities of produce and varieties of products offered for exchange from all points of the compass around Memphis. Trading areas from as far away as Uruk along the southern Euphrates River in Mesopotamia to the east, Nubia to the south, and all the distant oases scattered across the surrounding deserts wanted to establish commercial relationships with traders in Memphis.

If Senna could gather to himself the loyalties and a reputation for fair trading of these varied and diverse interests, his economic future, and that of his family would be assured for many years into the future.

====

"Building a boat capable of commercial travel on both the Upper Nile, and the Nile Delta is foolish, Father," Hesina said. "The shipping from growing areas in the Delta is too scattered and undeveloped, and the channels are too variable year-to-year. A boat built for shipping in the delta would never be fast enough or have enough capacity to compete with the big, fast boats built for shipping bulk grain on the Upper Nile.

"Far better to build two boats; one for each type of shipping." Hesina paused, "but it is even possible that any boat built for the Delta cannot earn enough to fatten the single owner no matter how well designed, and no matter how fit and eager the crew. Some kind of an association of shippers may be feasible, but it will take time to find enough warehousemen and shippers with enough uncommitted resources to make it work long enough to establish a value for the service."

Senna inclined his head in Hesina's direction, and chuckled. "As always, daughter, your insights into the problems of business and trade opportunities are as good as any in Memphis. I believe your thinking is correct, but I will need to think further on the idea of building a single boat to travel both waterways. I like the idea of forming an association of shippers to develop a system for getting produce and delivering supplies and equipment between the Delta farmers and the markets in Memphis.

"Once we establish the Delta service from Memphis, we can think about a warehouse and port on the Mediterranean Sea. But that is in the future. "For now, my daughter, I have another business proposition for you to consider. I have been aware of the jeweler from Uruk, and his flaunting of the Lapis Lazuli among those with refined tastes and plenty of money. He has been here before. He spreads precious fragments around like a monkey scattering his turds, and the activity is having an effect. Many are talking about it. If I am able to find a supplier, and a foothold in the market for it, I think we will gain much if we can bring some quantity of Lapis to Memphis."

"You know my interest in expanding our enterprise father, but you have not yet tried to expand into markets with sources of supply beyond the reach and protection of Pharaoh. How will you protect your Mesopotamian suppliers when they are trying to bring a shipment of Lapis Lazuli to you here in Memphis? I don't think your experience so far with your sales agent Laret in Jerusalem will be of much help. Besides, I am not sure I trust Laret. The quantities in his reports to us do not always add up, and the mistakes seem always to go to his advantage.

"I am well aware of Laret, and his loose connection with the reporting of his trades. To some extent I forgive him because his connection with hieroglyphics is also loose, and with hieratics almost impossible. Worse, he does not have much experience with the counting systems you have set up. But he performs other services for us that I do trust and cannot do without as long as I seek to trade our goods anywhere in Jericho, or anywhere along the Jordan River to the Sea of Galilee. As I have told you many times, I dream of expanding our enterprises into the lower Euphrates River in Mesopotamia, and Laret and Jericho are my stepping stones toward that goal."

"Yes, Father," Hesina replied.

After a pause to drink some beer, Senna continued with the topic he had wanted to discuss with Hesina at this meeting. "The Uruk jeweler and trader in Lapis is named Faiza. He will be returning to Uruk in southern Mesopotamia soon, and he has invited me to send one of my personal representatives with him to explore whatever markets we might have an interest in.

"I want you to be that representative. But you and I know that the venture will be filled with great danger and great discomfort. The rewards, if any, could be as smoke. They could blow away in the first breeze. Please give this some thought and speak to me about it in a month.

"In the meantime, I want you to go up river with Serenen, to check on the irrigation project that our friend, Heti, is working on at Beni Hasan. He is trying to increase the quantity of grain that he can plant and ship to us in the spring. I need your assurance that he is staying on schedule. If he is having trouble, I need to know what is causing it, and I need your ideas for resolution of any difficulties you might find.

"I know this is a lot to ask, my daughter, but I have many things going on, and I need your help in making sure everything is going smoothly."

"But father, I have never been upriver. How will I know what I am seeing?" Hesina asked.

"Serenen has been upriver many times, and he knows Heti very well. They worked together to prepare the fields that now produce so much good barley and beer for us. I want you to have the experience of traveling with Serenen because I think he will need to be one of your companions on your trip to the north with the Mesopotamian trader… if you decide to go, that is."

Hesina knew that her father expected her to go with the Mesopotamian jeweler, and that he would be greatly disappointed if she chose not to. She had to admit to herself that the dangers, discomforts, and difficulties of the trip gave her much concern. However, the opportunity for adventure and learning, and exploring new possibilities for enterprise meant that she would have trouble finding any reason for refusing to go that would not make Senna very angry.

"Yes, Father," Hesina said. "I'll tell Serenen to prepare so that we can be ready to go upriver in the morning."

"Talk to Serenen about the trip to Mesopotamia," Senna said.

"Yes, Father," Hesina replied.

=====

Early the next morning the sun rose on a stiff northerly breeze. Hesina understood that the breeze would mean a shorter trip up the river to Beni Hasan; possibly 5 days instead of 6. Serenen had assembled a stack of tools needed by the tenant farmers for the irrigation and planting on Senna's plot of land.

One of the Nile's many boatmen well-known to Senna and Hesina, Akar, was sailing up river with his crew of 10 oarsmen to the Upper Nile capital at Thebes. At Senna's suggestion Akar had invited Hesina to come along with whatever other staff and equipment she would need to take with her. He would drop them at Beni Hasan, then continue on his way upriver.

Akar had now come ashore at a dock close by Senna's warehouse to let Hesina know that he was ready to load their gear. Akar is a big man and strong. He moves toward Hesina with the confidence of one who is very close to Hesina, and to her family. "But first, my sweet beauty, you must serve me tea," Akar said.

"Of course, Akar, "Hesina replied with a broad smile and all her teeth showing. "Do you think you will have any problems finding your way upriver? You will be hauling precious cargo, as you know." Hesina let a smirk cross her face, then motioned to one of Senna's household slaves to bring some tea and honey-sweetened biscuits. "It looks windy this morning. Can you tell me how the trip will go?"

Akar responded: "It may be a bit rough because of the wind, but the disturbances in the river flow because of the summer flooding has now subsided. I have a very capable crew. With our sails out we should be able to capture most of the north wind. If we do a good job of sailing and rowing, we might be able to cut the time to less than 5 days."

"I'm sure you know, this will be my first time on an extended trip upriver. Are there any dangers on the river that I may not be aware of, and might not have prepared for?" Hesina asked.

"There is a lot of traffic moving up the river with tools for the irrigation and planting, as are you and Serenen. Have you made arrangements for pickup and return to Memphis?"

"Yes. We are allowing ourselves 5 days helping clear the irrigation channels at our fields in Beni Hasan, then my father has made arrangements with Kames to bring us back. Do you think Kames is a capable boatman?," Hesina asked.

As they talked a raven landed near the place where they sat. The bird hopped up as though expecting a piece of biscuit. Hesina noted that the raven had a single white feather on the top of its right wing. She tossed the bird a morsel, and the bird hunkered down to look up at Hesina as if to say: "Wow. Thanks." Hesina smiled and turned back to Akar.

"Well, Akar. Is Kames a good boatman, or not?"

"He is a good boatman if he can find a good crew. He usually can find such a crew, but that is never guaranteed coming from Thebes. Your man, Serenen, is a good man with either an oar, or a sail. I think he is fully capable of running a boat of his own if you and your father ever decide to put your own boat in the water.

"In any case, on your return to Memphis you will be traveling downriver with the current. That is a much easier task in terms of effort and navigation, but never forget just how dangerous the Nile can be when you are coming down river with the current. The river is vast and relentless. She doesn't care a fig about the humans who try to earn a living on her. Remember: things happen fast when you are in control. They happen much faster when you are out of control on the Nile."

Serenen had loaded the tools, equipment, and two weeks of foodstuffs onto Akar's boat with the help of Akar's crew. Serenen motioned to Hesina that It is time to go.

But Hesina was not yet ready to go. "My father is sending one more passenger to go with us. We need to wait for him. I have also asked my friend Nena to travel with us to keep me company. I hope this will not add too much of a burden to you and your crew, Akar."

"Of course not, my pretty one. I know your friend Nena. She is a very charming woman, and I know her to be among the circle of friends who are familiar with Pharaoh. You can count on my complete discretion. Who is the other passenger?" Akar asked.

"I believe the one coming up the street is our last passenger." She pointed to a man short in stature, but well-muscled and at ease on his feet even though his burden appeared to be heavy. "This is Paru, a lieutenant of guards in the palace of the Pharaoh. My father has asked Pharaoh to send him with us to protect us, and to assure our peaceful intent to any on the river who might inquire as to the Pharaoh's interest in our trip."

"Do you feel this extra protection is necessary, my beauty? Do you and your father not trust us enough?," asks Akar.

"Please put those concerns away, Akar. They are not worthy of you. My father wants Pharaoh to be fully aware of our plans for the property at Beni Hasan, and he has sent his most trusted, and most physically capable lieutenant with us. While we are underway, Paru can assist with the rowing and sail setting as they become necessary. He is under your command for anything to do with the operation of the boat on the water, even if there might be minor problems with people having hostile intent against us.

"If a minor problem escalates, however, I will want you to defer to Paru's experience in combat. If he calls on you, or any of your crew for assistance while we are under threat, I will expect you to give him all that he may need. You have seen the heavy bag he has brought on board. I can assure you that the bag contains weapons, and Paru is very well experienced with their use.

"To be clear. Paru is here primarily as a diplomat, and representative of Pharaoh. I do not expect trouble on this trip, but if trouble occurs, Paru will be in charge of our defense."

"Yes, my pretty one. You have my full support in this," Akar said.

"Thank you Akar. Here comes Nena. Once she is aboard with her things, we will be ready to sail."

=====

Nena is a tall and darkly attractive beauty, though, today, she is dressed modestly with a shawl to protect herself from the wind. Akar guided her to an area in the center of the boat. She and Hesina will be out of the way of the slaves who would row and set the rudder and sails as ordered by Akar from the back of the boat. The area had been closed off to give the two women some privacy. The men would sleep, when possible, near their assigned posts.

Akar's boat is designed for work so it's amenities were few, especially with Pharaoh's friends and agents as passengers. All aboard understood, however, that the lack of ostentation was purposeful. In that modest way, the boat would not call attention to itself, and risk becoming the target of river-borne thieves.

The intensity of the work day depended on which of the river's many moods were most in evidence, or most likely to change through the day. Akar always kept most of the rowers in position at the oars, ready to respond to his commands. Rowers not in position and ready to row were encouraged to rest near their assigned positions. They could eat and drink a little beer, but they were not to eat or drink to excess lest they develop cramps if called upon to row in adverse conditions.

All shared the basic food stocks of bread, beer, dried raisins and dates, and some salted meats. Some on the boat, though, like Pharaoh's friends, usually had a small pot of expensive honey for added flavor and pleasure.

If the wind was strong enough, and the sailors able to set good sail to catch it, the rowers might have an easy day. On the other hand, if the wind became adverse to their direction of travel, and the sails became useless, they might have to work doubly hard for the greater part of the day.

Some of Akar's crew were accomplished musicians with their own drums, tambourines, and lutes. Off-duty rowers often entertained themselves and everyone with their music and song.

Once darkness fell, the work day for the boat crew was over. The dangers of trying to travel on the powerful river when it is too dark to see are too great for anything but the most skilled crew motivated by the greatest of possible emergencies. Toward the end of the day, Akar would look along the shorelines for a place to park the boat so they would be safe from thieves, and from the other creatures of the night like crocodiles and hippos who might have too great an interest in the human and other contents of his boat.

Each of the rowers and sailors would have an assigned time during the night where they would have to keep watch for danger. When on watch each would keep time with a pot of water that dripped. When empty, the watcher would rouse the next watcher out of his restful sleep. Sometimes a watcher would not be able to keep his eyes open and would fall back to sleep while on watch. Any watcher who failed to stay awake on his watch could expect several lashes from Akar's whip.

Akar sympathized with the members of his crew on the matter of watches, and the difficulty of staying awake. He was fair in his enforcement of most rules of boat operations but failing to stay awake on watch was punishable with the whip. There were no exceptions.

If at dusk Akar could see a village or settlement along the shore line, and if he knew them to be friendly to boatmen, he would often put in, and offer to have the villagers prepare meals for those on the boat who had goods, or useful household items like cloth, to trade for a good meal, and a night's sleep on dry land.

=====

The arrival at Beni Hasan was uneventful, as was the time Hesina and Serenen and their entourage spent there. They worked with Heti and his family in the preparation of the fields for planting. The Nile flood this year had gone well, and with sufficient quantity that little effort was needed to assure that the standing waters were properly distributed. They group had enjoyed many pleasant evenings together.

It had turned out that Paru, in addition to be being an experienced warrior and Lieutenant of the Pharaoh's palace guard, was a poet of considerable talent. He was able to entertain everyone with vivid stories of great battles between the gods Osiris with his wife Isis, and Seti, leader of the underworld.

Hesina had always known how to work in the fields. Her father had not insulated her from the necessary work of his warehouse business. Now, after this visit, she would know the sore muscles and blisters that come with turning the fertile ground in order to produce quantities of barley and beer. She had also gotten over her shyness about nudity during the heat of the day, and usually stripped off all of her clothes as the sun rose higher in the sky.

Nena at first felt that she should not bring embarrassment to the house of Pharaoh by working in the fields, and, even worse, be seen naked in them. At one point, Hesina essentially shamed her into picking up a digging tool to help with clearing the irrigation ditches. Soon, Nena got into the habit of stripping naked in the hot sun.

Heti was delighted that such beautiful women would help in the fields and show off their womanly charms at the same time. He, his wife, and another older couple lived on the farm to do the planting and help fend off the creatures who would come to feed on the green shoots when they popped out of the ground. The other members of Heti's family, and their two friends also worked naked, but they looked away from the young women. They all had a note of sadness in their eyes, perhaps remembering their own younger days.

Akar the boatman had continued up river after dropping Hesina and her party off at Beni Hasan. He had thought that he would not return downriver for many days but told Hesina that he would send a message to Kames that he should be prepared to take Hesina and her group back to Memphis within the next several days. "However," Akar had told her, "if I cannot find Kames, or if he is not ready, I will come back downriver myself to pick you all up for the return to Memphis. I will plan on seeing you again in 5 days."

At mid-morning of the 6th day, Hesina looked up from her work in a ditch to see Akar's boat on the river coming toward her. She had another day of work to do in the fields, but that work would go quickly if Akar could persuade his boatmen to help. Once ashore, Akar took note of Hesina dressed only in a piece of cotton around her waist as she worked, and Nena with nothing at all. He knew he could have a problem with his boatmen if they were allowed to grow randy at the sight of women in such a state of undress. He told his men to put him ashore near the fields, but to take the boat a little way further down the river to the next copse of trees, then tie it up and tidy up the boat for the trip to Memphis.

"Tie the boat up near the trees, then stay out of the heat for the afternoon. By late afternoon, when it is cooler, bring the boat back up the river to these fields. Maybe we will get a good meal tonight."

As his boat departed, Akar approached the two women. After running his eyes hungrily up and down their sun-darkened bodies he grew concerned about the swelling he felt under his waist cloth. It would give Hesina too much satisfaction to allow a complete erection to develop. She should not know that she had such an effect on him. He looked away for a moment, and then he finally spoke: "Well, my two beauties. Hard manual labor becomes both of you."

"Be careful of your language, Akar. Do not take advantage of my need for comfort in this heat," Hesina remarked. "How was your trip upriver? I am very glad to see you back here by the way."

"The trip was good, my pretty one, but Kames was nowhere to be found, and my own plans were thwarted when the men I intended to meet with could not be found. I will be delighted to take you and your party back downriver, but I need to know how much more time you need in the fields." Hesina replied. "If Nena and I and the families here are the only ones working then it will take us another full day. On the other hand, however, if you can release your boat crew to us, we can probably finish in half a day." Akar paused and looked down at his feet for a moment. "I will make you this offer, pretty one. If you and Nena will put your clothes on while my men are working with you, we can probably get all the work done. I must admit to you that the sight of both of you women naked and being very physical in your work can give a man ideas that he should not have if he values his friendships. You should know that my men are less discrete than I am. I can enforce a dress rule on the boat, but You cannot temp them too much by flaunting your bodies around them. Do you understand me?"

"Yes I do, Akar. I understand completely, as does Nena. We will wear our waist cloths and breast coverings while we work with your men, and we will try not to temp them unnecessarily on the trip back to Memphis. We will rely on your ability to keep your men calm around us.

"Now for more important things, I have alerted Heti to your arrival, and have asked him to prepare food for this evening's dinner. We have discovered that Paru is a poet of considerable ability. If those of your men who can play instruments will play for us tonight, it will be a joyous event.

"For the very reasons you have expressed concerns about, we have decided to distribute our small stock of beer to some of the neighboring farmers. So. There is no beer. Your men will have to be satisfied with music and poetry."

Akar laughed. Hesina and Nena put on their modesty cloths, then turned and began to walk back to the fields. As he watched them walk away, he noticed that their swaying hips were barely hidden by the skimpy cloth. Watching them made Akar wonder if it was not a better plan to have them naked and looking more like little girls in their bath rather than grown women who are the objects of lust and imagination by grown men.

Akar moved up beside Hesina and asked for a private moment with her.

"Yes, Akar. What is it?" she asked.

"Where are Serenen and Paru?"

"They are working the fields," Hesina answered.

"I think we need to make sure they both understand how I see the problem of these boatmen and the casual attitudes toward nudity that they have, and that you and Nena have, and what you and I have agreed for the return trip to Memphis."

"I think you are overly concerned, but I will ask Serenen and Paru to join us for a few minutes before we return to work." Hesina turned toward the two men.

Akar waited. As Hesina walked toward him with the two men, he could see the concern in their conversation. He noted that both men were naked.

Paru greeted Akar. "Did you have a good trip upriver? Paru asked. "Any crocodiles?"

"No. No crocodiles. A few hippopotamus only. They stayed well clear of the boat."

"Good. Hesina tells me you are concerned about our return to Memphis. Something about nudity."

"Yes. Something about nudity. It is a custom in the rural areas for Egyptian farmers to work without clothes in the heat of the sun. Hesina and Nena have also picked up that habit and seem to enjoy the feeling of freedom. The men on my boat also work without clothes when the sun is hot. My concern is that the nudity is bad when there are two such attractive women on the boat, who also enjoy nudity, and seem not to worry about the effect of it on men.

"As captain, any problems that might arise from the temptations that could overtake some of the men, could put the boat itself in jeopardy.

"I want to require that everyone cover their nudity on the boat while we are underway, no matter how hot it might get. That will include you, Paru, and Serenen, and it will especially include Nena and Hesina. Hesina has already agreed, and I hope that Nena will also agree, but I need the support of you two men if I am to enforce the rule.

"I believe it will be very difficult to enforce the rule among at least a couple of my rowers. Since I need to be in full control of the boat and the rowing at all times when we are running downriver, I will have to have one or both of you ready to apply whatever punishment might be needed to calm any bad situations. Do I have your agreement on that?" Akar asks.

"Yes. Of course. "But you know your men, and you know the potential trouble-makers. How will you communicate your concern to us, and how will you show us the problem so that we may act on it?" Paru asks.

"I will tell my crew what I want them to do to cover their nakedness, and I will tell them what I have in mind for punishment if any are unable to control their passions… or their erections. They know me to have a strong whip arm, and they know that if I warn them about expected behavior, that I will use the whip on violators.

"How about you, Serenen? You have not yet spoken in my presence. What are your thoughts?"

"Should a problem arise, Captain, I will defer to Paru, unless the problem is urgent and I am closer to it. If I must deal effectively and quickly with the problem, and Paru is not able to assist, I may heave the offender overboard, and take over his responsibilities on the oar, or rudder, or sail. I am well-experienced in all aspects of boat operations.

"Please understand, Captain, that I have been attendant, protector, and mentor to Hesina since she was a small child. I take my responsibilities very seriously."

Paru spoke up. "Be assured of my full confidence in the words of Serenen, Captain. Before he became protector to Hesina he served with me in the Pharaoh's personal guard unit in the palace. We have been involved in many battles. I trust him with my life."

Akar sat back and rubbed his chin as he thought about the words of the two naked men standing in front of him.

Nena stepped forward. "You have not asked us what we think of all this, Captain. I cannot speak for the sexual experience of Hesina, though I suspect she has already learned much about the journey of life, even on this, her first trip of some distance upriver. Speaking for myself, I have already learned much of the world, and of the base interests of most men in their imaginings about sexual congress, particularly with women as beautiful as Hesina and myself.

"I imagine we could drive some of them crazy with desire if we were to remove these modesty cloths from our bodies. If we show some extra flesh from time to time, it may help us all figure who might need more careful observation. We will work with you on this. As women, we will both be at very great risk for physical harm, and we have a great deal at stake in the success of this trip. We want to avoid the problems you speak of as much as you do."

Hesina nodded her assent.

Akar speaks. "You have all given me great relief concerning this problem. I thank you. In the morning while everyone else is working in the fields, I will want to work with Paru to correct any possible sources of problems on the boat itself. Though I will be moving the boat toward Memphis with haste, there will be several long hot days on the river. The women's shed will need to be completely shielded from casual views by the boatmen. At the same time, to avoid boredom, the women will need to have an opening toward the shore for their viewing, and there must be an opening for breeze. I will assign the boatmen in such a way that they won't be constantly thinking of what the women in the shed might be up to; …but there will be leisure times when we are under sail, and they might all be unoccupied, and left to their own imaginings. That is the time we must be most alert. Two of my boatmen are particularly trustworthy in enforcing my wishes among the others. I will ask your advice, Paru, on how they should be deployed, and how advised as to their responsibilities.

"Let us now see what we can do to finish this day's work in the fields, and then prepare for the evening meal," Akar concluded.

====

The food was good but the party was quiet and didn't last much beyond the last bowl of honey-sweetened dates. All the participants were tired after several days of hard work in the fields and on the boat, and another hard day of work would begin again the next morning.

The boatmen were not happy with the need to wear waist cloths while working in the fields, but most agreed, with varying degrees of acceptance, under threat of being put off the boat, to the need to respect the women's modesty. Those most sullen and reserved in their acceptance were noted by Akar, and by each member of Hesina's party.

The work in the fields progressed quickly through the morning to completion as the sun passed its highest point in the sky. After a lunch of bread and a small ration of beer, they were able to finish loading the boat with tools, weapons, and foodstuffs. Akar wanted to get as much time as possible to move down the river before dark, so he pushed everyone to move quickly to finish the loading. Once done the boatmen cast off from the small dock, and into the Nile's steady, and inexorable push downriver toward Memphis and the great Mediterranean Sea beyond.

Once on the river's main channel, the oarsmen began their routine of coordinated rowing designed to give Akar maximum control over his boat as he searched the river's surface for the fastest currents. The rudder men and sailors were tuned to Akar's every hand gesture and to the nuance of his every verbal command. They knew exactly how to set the sails and rudder when commanded once he had given it.

As well-experienced boatmen, Paru and Serenen were alert to the same signs given off by the river in its course and could usually anticipate Akar's command. Even so, they delayed their intuitive actions by split seconds in order to give Akar that time to accept or reject the moves he knew they were about to make. Akar had the more general view of what would need to be done – perhaps immediately - to maintain a steady and fast course across the water.

Paru and Serenen had the greatest respect for Akar's skill as Captain. Watching Akar work was like watching a man supremely confident of his skill and his crew, and all were being tested. His actions were more those they imagined of a God than a man. They would do all in their power to support his effort to give his passengers a safe, and fast, trip to Memphis.

At the end of the day, Akar complimented all on the crew for their good, strong work through the day. As they tied up for the night, Akar told his men that they would have to sleep ashore in the trees, the better to watch for the various creatures, or persons, that might otherwise threaten the boat, and our precious cargo.

Akar spoke. "I regret that I cannot offer beer to you this evening because there is none on the boat. You will have to quench your thirst with river water, but I am told that this place has a source of very fine water, and I am sure you will be most pleased. There are extra rations of dates, salted meat, and honey to repay you for your work today. I'm hopeful that the weather and wind will hold, and we will have a good day tomorrow and the next day."

The Captain, Paru, and Serenen laid out their sleeping mats around the women's shelter. They inquired of the women, how their day had gone. Hesina said they had had a fine day, but wondered if, tomorrow, they might spend more time out of the shelter… perhaps even take a turn rowing, or helping to set the sail, or the rudder. "If we spend another day in pampered indolence, I think we will grow fat, and our moods will be unattractive."

"I am less worried about unattractive moods than I am about restless and bored boatmen imagining ways of relieving their sexual stress," Akar said. "I am mindful of your concerns, however, and will give it some thought. I'll let you know in the morning. Now it is time to sleep. Rest well in the encircling arms of Horus."

"And you as well, Akar. Thank you for your good work today," Hesina said.

====

Before dawn the next morning, Akar asked his distinguished guests to join with him to discuss Hesina's request of the previous day. "I have given much thought to this request," Akar said to all of them. "After watching them work in the fields, I can see that both of our women have as much strength, focus, and ability as most of the boatman, if not their experience and training. I can also see considerable value in having two more crew in the rotation between work and rest. However, I remain mindful and concerned about the possibility of too much sexual imagining combined with too little sexual restraint under some circumstances.

"I am willing to try to work you both into working positions on the boat, but you must remain as fully clothed as you can be, no matter how hot it is. Also, you must work in positions behind the men. I ask you both to understand how a man's sexual imaginings will rise if he is constantly watching a beautiful woman bend to her tasks in front of him. Do I have your agreement on these conditions, so far?" Both

women nodded their assent.

Akar continued. "In the back of the boat where the rudder men work I want Hesina to take her position on the right side of the boat. There are two rudders there, and I will want you to stand by the rear rudder. Both the rudder men on the right side have a sexual preference for men, so I see less risk of the kind I am most concerned about, though there is always the possibility that they could change their minds, if the women are as attractive as the two of you."

Akar turned to Serenen. "I want you, Serenen, to continue to support the work of the rudder men as you were doing yesterday. While there, always be ready to protect your charge from both sexual problems, and the kind of problems that can occur on a boat when people are not experienced with a ship's shifting movements when underway. The added problem of her inexperience with a moving deck while doing her work worries me a great deal, but I think it is manageable if you will maintain high vigilance. Can you do that, Serenen?"

"Yes, Captain. I will make myself ready to handle any event, and any circumstance that might bring harm to Hesina," Serenen said.

"What will you do if she goes overboard?" Akar asks.

"Hesina is a very capable swimmer. I am sure she can take care of herself in the water. Even so I will go in after her to assist, unless there is some more pressing business occurring on the boat. I am confident that Hesina can maintain herself in these waters long enough for you to get the boat turned around to pick her up."

Akar continued. "Good. Nena. I see you and Paru in the same basic position with the rudder men, initially. You will take those positions when I call a rest period for Hesina and Serenen. Do you have any concerns with what I have said? I should add that Paru will not go with you to the shelter. He and Serenen will trade places and continue to support the work of the boatmen."

Paru spoke for both when he spoke. "Nena and I believe your plan to be a wise one. To your question about what I would do if Neva were to fall overboard, my answer is the same as that given by Serenen. Both these women are very capable of taking care of themselves in deep water. They can certainly keep themselves afloat long enough for you to turn the boat around to pick them up."

"Good to hear," Akar replied. "Now let us eat so we can get underway as soon as there is enough light for me to read the river. We will try this plan for some part of the day, today, and we will talk about the results and possible changes this evening." He turned to the two women. "Assure me that if I say to you Hesina, or to you, Nena, that you must return immediately to the shelter, you will do so immediately, and with absolutely no nasty looks, and no comments of any kind.

Hesina and Neva replied together. "Yes, Captain." Hesina continued. "We can assure you of our full cooperation, and of our gratitude for this opportunity to be useful."

=====

The day's sailing downriver proceeded well. Hesina and Nena worked hard, and were very modest in dress, and in the exercise of their physicality while doing their tasks. Akar never felt the need to correct them, or to remove from their assignments prematurely.

The two homosexual rudder men quickly got into the spirit of the changes and were happy to let the women take over their jobs. They gave the women liberal tips on how they could improve their mastery of the procedures involved in changing rudder settings.

Akar, Paru, and Serenen were very pleased with themselves for making such wise choices in deploying the women. The women themselves were tired, but very happy that they had performed well.

Akar allowed himself to have a few pleasant moments about the day's work, but he held his good feelings in reserve over the probability of serious problems during at least three more days of travel. Before they would pull up to Senna's dock in Memphis, there were pockets of crocodiles and hippos in some of the sloughs ahead. Also, and worse, though he had never had to deal with them, he understood there to be a nest of boat-borne thieves at work on various parts of the river at various times of the year.

He had heard no talk of these thieves since the annual flooding had subsided, but the isolated reports of their return to action soon was very much on his mind. The kind of fear generated by these reports, and the preparations needed to counteract the hostile acts that were the subject matter of the reports was the business of Pharaoh. Akar resolved to discuss these concerns with Pharaoh's representative on his boat, Paru, later.

Over a meal of bread, salted meat and lentils, a small ration of beer, and dates for dessert, Akar whispered in Paru's ear that he would like to talk with him away from the others. After they had settled some distance away, each with a cup of beer, Akar spoke. "When I was upriver some of my fellow boatmen were talking about thieves active on the river in this area. I think Pharaoh would be aware of thieves if they were present and would have a plan for dealing with them. Do you know anything about thieves here, or about any plans for dealing with them that Pharaoh may have in mind?"

Paru replied. "Pharaoh is aware of thieves working in this area, but we have not had any reports of their activity last year, or this year. Their last attack was overland two years ago in the Delta north of Memphis. They met stiff resistance from the farmer's down there and we have believed that they were considerably weakened by death and injury. Pharaoh has an active interest in the thieves but has not yet set up any focused group to deal with them in any organized way.

"Part of the reason for sending me on this trip is Pharaoh's interest in my report to him on my return. So far, I have nothing to report regarding thieves. Do you have any reason to believe that we need to increase our awareness, or perhaps our weaponry, Akar?"

"I am not sure. The boatmen on the river above Beni Hasan, who passed these rumors to me, had not seen any incidents, as you say, this year, or last. I have feeling, though that we need to become more aware in general, and more alert to possible threats. We also need to assess our stock of weapons, and perhaps add more."

Serenen joined the two men. Paru spoke. "I am glad you are here Serenen. Akar and I have been discussing the possibility of thieves. I think we need to make more spears and arrows. The spears should be stout enough to fend them off if they have boats and try to ram us. I have material with me for stringing more bows, and for binding rocks to branches to use as clubs. Each boatman should have his own bow, supply of arrows, and club. I have two short knives, and one long sword."

"What about the women? Serenen asked. If we come under attack they may make us a more desirable target. Perhaps we should keep them in their shelters until we get to Memphis."

Paru responded. "No. They are both capable of being effective soldiers and sailors in my view. If trouble comes our way, we will need every hand on deck and armed. I think it is important to keep them on deck learning as much as they can about rowing, and about setting rudder and sail. I won't advocate that you relax the modesty rule, nor your strong enforcement of it, Akar, but once you advise your crew that we will be gearing and preparing additional weapons for possible attack by thieves, they will be much more concerned about being on the lookout for attack than about keeping their erections under control so they won't have to feel the Captain's whip."

Paru continued. "Except for tonight's rotation of guards, the rest of us should get as much sleep as possible. Tomorrow everybody will need to get up before light to disperse and start searching for weapons materials. Everyone, including the women, needs to make his and her own weapons. I will help each of them to build a stack of weapons that are reliable and effective. We will need to distribute the weapons around the boat in a way that an approaching boat will not see them, but each of our crew will be able to get to them quickly if the boat has hostile intent. I think that job is best left to you and your most trusted boatmen, Akar." "You

can count on it, my friend." Akar replied.

=====

After making and loading the weapons into Akar's hiding places on the boat, the boatmen and Akar's guests were not able to get underway until the sun had reached its peak in the sky. The rest of the day was mostly uneventful. The two exceptions occurred because the two women were now moving freely from rowing position to rudder position, to sail setting position through the day. Akar had not relaxed his modesty rule, but the sight of two such beautiful women moving quickly among the men, and occasionally bumping into them when the deck shifted required more discretion than two of the boatmen were capable of maintaining.

Akar applied his whip judiciously, to the men as their erections popped up as a result of inappropriate – and possibly unavoidable – touching. Once touched by the whip, the shame-faced offender did not allow the problem

to occur again.

The next two sailing days – their third and fourth on the river - were also uneventful, but, late on the fourth day, Akar felt and smelled a change in the air. His long experience on the river told him that the change was not good.

As he looked upwind to the west, he saw that a sandstorm was building on the Western horizon. The churning dust cloud was already far higher than the tallest buildings or mountains that Akar had ever seen in his life. Such a storm this close meant that they would have to steer for shore and tie up immediately until the storm passed.

As the boat moved closer to the river's west bank, Akar saw an opening into a slough that could offer more protection from the storm. It would require fast action by the crew to collapse the sails and modify their speed and direction in a way to get into the slough without crashing into the bank.

Such a crash at their current speed could wreck the boat, and possibly sink it.

Akar shouted his orders to his most trusted, and most capable crew.

Serenen and Paru both heard and saw the urgency in Akar's voice and they moved quickly toward Hesina and Nena to get them to lay down on the deck out of the way. The maneuver's success depends completely on correct actions by the most experienced of Akar's crew.

Paru realized intuitively that the sight of the women's bare flesh as their modesty cloths slipped during their scramble to get out of the way could cause an accident, even a man overboard, under the pressure of need for such precise maneuvers, but all seemed to go well.

Akar ordered the boat turned to the bank as soon as they entered the slough. They hit hard, and those who were not quite ready for the rough stop were thrown to the deck or thrown overboard onto the shore. One man went into the water but was able to quickly scramble back out again.

Akar ordered the boat secured to the shore and secured against the wind and sand. Akar knew that if enough sand found its way into the cracks around the hull and the deck, the boat could well sink. Worse, if they had to find their way out of the slough, the sand might weigh them down enough that they will not be able to clear the slough's storm-shifted sand bars.

Once the boat was secured, Akar and Paru walked ahead to look at the slough. They knew the problems of pushing the boat back out of the slough would be very difficult under the best of circumstances unless the slough led back into the main channel of the Nile. This turned out to be the case, but it was at least 40 paces to the end, and they would have to clear portions of the banks to fit the boat through it.

It might also be variable depths in the slough's bottom that will need to be cleared. Akar jumped into the water and began to probe with one of the spears. He knew his boat, when loaded with cargo, needed a draft depth about as high as his waist. As he probed, he found one hump that would need to be cleared. In general, Akar felt that they could move the boat forward to the end of the slough in about a half day.

Paru spoke. "We need to get out of the water. There are probably crocodiles in the swampy area beyond the banks of this slough. If they are nearby, it won't take them long to find us." After a pause, Paru spoke again. "Perhaps we need to move the boat further down the slough now while we have plenty of daylight even though it will be stormy and miserable to work in the middle of a sand storm. Better that than fight with crocs while we try to move the boat in the morning."

"That is good thinking," Akar said. "I'll talk to the boatmen. We will need to post guards with spears to watch for crocs, and we will need to distribute our digging tools to the strongest diggers. The women can keep watch from the boat. They should be prepared to climb the masts from time to time so they can get a better view."

Akar continued with Paru. "You and I need to probe to the end of the slough to make sure there are no surprises under water."

The two men walked back to the gathered boatmen to explain their plan. After some time, tools and spears were distributed, and the men bent to the task of widening the banks and watching for crocs. Four of the men accompanied Akar and Paru downstream in the slough to clear any underwater mounds that might hinder the boat's passage.

Hesina and Nena used ropes to rig a way to climb quickly up and down the boat's mast. Hesina took the first watch near the top of the mast. Nena moved to the back of the boat to watch for possible threats that might be coming from the river.

From Hesina's perch, the horizon looked clear of crocs, or other threats for as far as she could see, but the storm was coming closer, and she could not see beyond the storm front. She knew she would have to climb down from the mast when the storm's front got close to the boat because she had no idea what the force of the winds, and the blasts of sand could do. She would call out when the storm got close so the crew on the shore and in the water could take cover. She would not leave her post on the mast until all were prepared to withstand the storm's fury.

The digging work was going well enough that Akar had to decide whether or not to free the boat from its temporary moorings in order to advance the boat as far as the digging and clearing had made possible. Once moved forward, he would have to re-secure the boat to ride out the storm. These would all take time. He decided he needed to take a look at the storm's progress before deciding to free the boat from the shore.

"Come down from the mast, my pretty one. I need to see what the storm is doing." Hesina came down the mast. She had tried to protect her modesty, but urgency and difficulty of what they were trying to do, made the protection of modesty too much. Akar was not much in the mood for modesty enforcement either, since he had long since given up trying to keep his waist cloth in place.

From the top of the mast, Akar noted that only one or two of the boatmen still had waist cloths. It was simply no longer important given the urgent need to deal with the approaching storm. He looked carefully at the storm front. He realized he probably had just enough time to move the boat perhaps 20 paces forward and re-secure it before the storm hit them. He called his two most trusted boatmen to his side. He told them his plan, and the need to get the other men to hurry back to the boat to help move it.

Before descending from the mast Akar noted that the storm front was already crossing the river about a half day's sailing to the north. He allowed himself to have a moment of optimism as he recognized that there was a chance the storm would miss them, or at least only brush over them without causing any serious problems. He said nothing of this to anyone.

The storm might also make a sudden change in direction that could bring the worst of the winds and sand-blasting over this part of the river. He needed to keep everyone on high alert, and well-focused on the problem of clearing the slough of obstacles to the boat's continued passage through it.

Nena took a turn at the top of the mast. Hesina moved to the rear of the boat. So far, there was nothing to report. Some boats had pulled to the shore above the mouth of the slough where their boat now sat. She did not see any threats among them, though they were too far away to say for sure.

She hoped that after the storm these fellow river-boatmen would be willing to assist with any problems the storm might cause for them.

Nena shouted out to Hesina that the storm was very close now. Hesina ran forward to alert Akar. The boatmen would have to now stop digging and focus on securing the boat to the shore. Nena was pleased to see that, by their digging, they had advanced the boat to a place within sight of the slough's opening back to the main river.

=====

Akar, his guests, and his crew of boatmen did escape the worst of the storm. There was wind and blowing sand through much of the night, but no damage to the boat; though quantities of sand had to be moved overboard. Because of the possibility of crocodiles, Akar ordered everybody to sleep on the boat deck while at least two men were kept on watch for them through the night. Everyone covered their faces to keep the sand out of their nostrils and mouths as best they could by using their modesty cloths. Akar's concern about too much sexual imagining because of the women on board was not a concern while the entire crew sought rest from the day's intense labors.

As soon as there was enough morning light to see beyond the boat, Akar climbed to the top of the mast. He looked around the boat and did not see any threats. He told the two watchmen on duty to get everybody up to begin digging so they could free the boat and get it back to the main channel as soon as possible. With a good day's rowing, Akar believed he could get the boat within sight of Memphis by dark this day.

He felt blessed by his success in bringing his boat and crew through all these difficulties. By the next morning he hoped to deliver his guests back to the safety of their homes, and to the enfolding arms of the great god Horus.

Nena was awake and moving about the boat to take some food before beginning her turn at the top of the mast. Suddenly there was whistling in the air and a sharp cry from Nena. She collapsed to the deck with an arrow through her thigh. Paru moved immediately to his cache of weapons. He motioned everybody on deck to stay down and get out their own weapons from their hiding places without showing them until Paru issued his orders. Paru raised his head up until he could see and assess the threat. Another boat was moving in the main channel. Two archers were preparing to shoot at targets of opportunity on Akar's boat. Paru determined that they were not in a good position to capture a boat and take it as a prize. They were moving too fast down the river. More importantly, the boat's captain was focused on his boat, not on any possible shore-bound prizes. Paru advised his two best archers to arm themselves then wait for his command. On command, they were to rise with him, to aim carefully, and to shoot in the direction he was indicating by his hand.

On Paru's command the two archers rose and loosed their arrows in the direction he indicated. All three arrows found marks, and the two archers on the hostile boat fell. None took their place.

Paru and Hesina moved to Nena's side. Serenen brought a knife to cut the arrow away, and a cloth to bind her wound and stop the bleeding. Paru was not yet sure whether the arrow had hit one of the vessels in the leg that carried a lot of blood, but he did know that Serenen was the most experienced healer on the boat. Paru motioned to Akar to get the men working on freeing the boat; that Nena's life may depend on how fast he could get them to Memphis.

All Akar's guests and crew now bent their backs to the task of freeing the boat, and rowing as fast as they were able to get to Memphis before nightfall.

====

Akar pulled his boat, his guests, and his crew up to the dock in front of Senna's properties early the next morning. Paru ran as fast as he could to find Senna and look for a priest or physician to look after Nena's wounds. Despite the best efforts of Hesina and Serenen to stop the flow of blood by applying a mud poultice, they were concerned that they had not done enough. They tried to keep her warm, but she had gone into shock from blood loss. They feared for her life.

Paru soon returned with Senna. He hugged his daughter and expressed his great concern for her welfare following the sand storm. His voice was shaky as he asked if she was OK. Hesina was very glad to see her father again. Even though naked, she gave her father a great hug.

Her mother, Salama, arrived on the scene with cloth to cover her daughter in an attempt to restore her modesty. Hesina had not thought much of her modesty since they pulled into the slough ahead of the sand storm, but she accepted her mother's cloth, and her mother's wrapping of it around her body gratefully.

Hesina knew she would need days to try to grasp and absorb all that she had learned on her trip upriver. For now she was grateful beyond belief for the love of her parents.

Senna cautioned Paru against bringing a Priest to look after Nena's wounds. "They are too enamored of their knowledge of useless potions and treatments that they think cause healing. Actually they are worse than useless. To the extent that they interfere with a careful analysis of what the actual disease, or health condition, or wound is, their application can lead to worsening of whatever condition is being treated.

"I believe the best treatment now is to have Serenen and Hesina continue with her care here. There is no need to move her further. Nena is a strong woman, and I believe she will recover from this wound. As always Serenen has brought the correct healing practice to her, and I believe she will survive."

====

Hesina found out from her father that the Uruk trader in Lapis had to leave sooner than he had hoped because of a shipping problem near the ancient city of Jericho. He begged forgiveness of Senna that he had not been able to wait for the return of Hesina but promised to return to Memphis next year. He hoped they could begin again their plan to take Senna's representatives to Uruk to investigate trading opportunities.

As a token of his commitment, the trader left a sizeable piece of Lapis to tickle the imagination of Hesina and Senna while they awaited his return.

With the constant care of Hesina and Serenen, Nena did survive her wound. She was soon restored to her earlier position as one of Pharaoh's most valued and trusted household attendants.

PANEL 3 - EMPIRE

Aello, a young woman, is a skilled armorer with the Army of Alexander the Great of Macedon. Alexander is moving his army of 50,000 into southwest Afghanistan near present day Khandahar. Aello sees the rugged peaks of the Hindu Kush mountains, far to the northeast, shrouded in dark storm clouds.

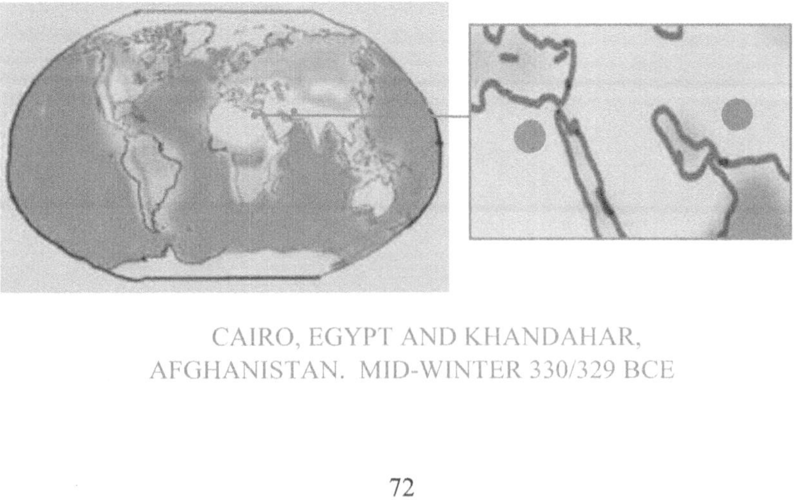

CAIRO, EGYPT AND KHANDAHAR, AFGHANISTAN. MID-WINTER 330/329 BCE

PROTAGONIST

Aello - A bright young woman who has skills and experience in logistics and organization, and in the making of weapons and body armor for Alexander's army. In Greek mythology Aello is the name of an Amazon warrior who wielded a double-axe.

Barak - A Greek of Jewish heritage; The father of Aello. A reluctant armorer to the Persians during their occupation of Egypt; a secret armorer to the army of Alexander while he fought Egypt's Persian occupiers; a happy convert to Alexander's side after the Persian defeat, abandonment of their garrisons in Egypt, and their army's retreat to the east.

Nikola - Brother of Aello.

Iva - Mother of Aello.

Tabu - Barak's steel maker. He is a very large, very well-muscled fellow; very important characteristics in making steel because of the constant repetition of the heating, shaping and pounding, cooling, re-heating, shaping and pounding, and cooling needed to make steel.

ANTAGONIST(S)

Alexander the Great – (b. 356 BCE – d. 323 BCE) Leader of an invading Greek army whose strategic purpose was the defeat of the Persian empire under Darius III. In those days, Persia included most of what we know today as the Near and Middle East from Turkey, Macedonia, and Egypt to India. Alexander's invasion was repayment for the Persian invasion and occupation of Alexander's home territory of Macedonia by the Persian Xerxes/(Darius1?) I in the 5th century BCE. Alexander was about 5' 6" tall. About average for the times, but very well built, very athletic, and very quick on his feet. At the time of this story, Alexander was 27 years old.

Kallias - one of two of Alexander's Lieutenants charged with overseeing the quality of the work of contractors like Aello and Barack. Kallias and Alexander often engaged in swordplay as a way of testing weapons. Sometimes the swordplay got a little rough.

Petros - Also charged with overseeing the quality of the work of Aello. Petros reports to Kallias. Petros was multi-lingual. He was invaluable in translating between the representatives of Alexander's polyglot armies.

LESSER CHARACTERS

Dinocrates – (b. ? - d. 278 BCE) chief Architect to Alexander the Great. …with him when he laid out the City of Alexandria.

MINOR CHARACTERS

Pelagios - an engraver referred by a scribe friend of Aello to carve a Greek saying onto a sword forged by Tabu for the Pharaoh Alexander.

Earlier in the Winter of 330 BCE, Alexander III of Macedon, with his army of 50,000 infantry, mounted horsemen, and camp followers, continues his march east from Egypt. He has been named Pharaoh by the Egyptians. He has defeated his Persian enemy, Darius III, and has opened the way to the Persian capital of Persepolis.

Alexander continues moving north to the Caspian Sea, then eastward toward Afghanistan. He enters Afghanistan near present-day Herat.

Alexander the Great is in the fifth year of his campaign to retake all the territories held by the Persians, from Macedonia and Egypt in the west, to India in the east. Alexander moves south and settles his army into a winter camp at the southwest tip of the Hindu Kush mountains of Afghanistan, an area now known as the Helmand River Valley. He will form a permanent military garrison near present day Khandahar. Nearby, the Helmand River flows southwestward from the Hindu Kush mountains in the northeast.

There are divisions in his army between the older Macedonians loyal to his deceased father, Phillip II, and the younger troops loyal to Alexander. The complications of conducting war and feeding an army in hostile territory far from home are further complicated by persistent rumors of a plot against Alexander, in part because of his reveling in his new roles as Egyptian Pharaoh and King of Persia. Resolution of these conflicts occupies much of his time and attention.

After a period of time to settle grievances and rebuild his army, Alexander will continue his campaign to conquer all of the remaining Persian territories as far as India. First, he will move his army along the Helmand River northwest to Gandara, an area at the foot of the pass across the Hindu Kush mountains in an area now known as Afghanistan's capital at Kabul.

As the Helmand courses along the northern edge of the Desert of Death, the eye reaches across the high desert dunes for the far horizon in the south.

It is spring. Kallias, an officer on Alexander's senior campaign staff –

the Companions – had been assigned to inspect the terrain and grain production along the fertile Helmand River. With Kallias' report in hand Alexander has decided to set up a garrison for those who will supply the tons of food, goods, armaments, and medical services needed every day by his Army. Kallias would lead the reconnaissance.

With Kallias was Petros, a life-long friend and assistant to Kallias in his duties as assigned by members of Alexander's senior staff. During the long and troubled march from Persepolis, Alexander has come to favor the two for their hard work, their dedication to both the work ethic, and the resolution of the ugly details of their difficult assignments. Most important of all these were their demonstrated loyalties to the ultimate objectives of the Army during the many threats of leaving to go home by the Macedonians, unhappy with the Persian coloration taken on by their Macedonian leader.

Today, the first day following Alexander's order to halt the march for resupply, Kallias has asked Petros to work with all the officers to help them take inventory of the health, fighting capability, condition of the weapons of each of the Army's soldiers under their respective commands. Though he is not to talk about it, Kallias also asks, on orders from Alexander, that Petros assess the attitudes of each of the commanders to determine their fitness for the battles and the long marches ahead as they push across the Hindu Kush mountains into northern Afghanistan.

These kinds of assessments were a specialty of Petros. He spoke several of the languages and dialects typical of Alexander's polyglot armies. Though he looked like a Greek Lieutenant high in the ranks of Alexander's chain of command, he could walk through an encampment of Thracians, or Spartans, or Athenians, or Jews, without suspicion, and pick up the detail and nuance of any cooking fire conversation.

======

On the march from Egypt to Afghanistan, Barak has become a principal armorer for the army of Alexander. He has a large crew of workers with talents, skills, and experience in metal forging, wood and leather cutting, and assembly of all the war materials, armor, and armaments needed by an army on the march. Barak is a Jew born of Greek parents in Egypt. He had, for many years, owned a small metal-working shop in the City of Memphis on the west bank of the Nile river in Egypt. The Persian occupiers of the time occasionally required the production of weapons and other war materials from Barak's shop.

The penalty for failing to produce such weapons when demanded was interference in his other contracts, and his services to other customers. Other coercive methods included the implied threat of torture and possible death to himself or his family. These were the typical tactics used by the Persians, and Barak's development of black market channels to the Macedonian army was the inevitable result.

Barak's metal-smiths had mastered the production of such weapons for the Persians, though he did save his best work for the weapons and armaments that he could deliver to the Greek army through the black market channel that had developed when Alexander's success in following in the footsteps of his deceased father, Philip II, in pushing the Persians out of Greece and Asia Minor became apparent 2 years before.

Alexander took note of the high quality of the products of Barak's metal shops. Some units of his army had acquired weapons stolen or smuggled from the Persian occupiers and made available through black market channels.

Barak's weapons had been used very effectively by his army during the seven month siege of Tyre, and the following battle of Gaza, both of which had threatened to block Alexander's way to his objectives in Egypt. After hearing of Alexander's siege tactics, and the battle with the Palestinians in Gaza, the Persian king in Egypt, Darius III, fled Alexander by running away to the Persian strongholds in the east and north, beyond the Sinai. Darius left Egypt without a fight.

After 200 years of oppression, the people of Egypt were so pleased to be rid of the petty tyrant, Darius III, that they welcomed Alexander and all of his army. When Alexander arrived in the capital city of Memphis in late October, 332 BCE, the Egyptian people anointed him as Pharaoh in November 332 BCE. The culmination of this celebration was his naming as 'Son of the Gods' by the high priest of Egypt.

Alexander believed this reception, and the glory heaped upon him and his army, to be a good omen that would favor him in his war to take his conquering army across all of Persia as far as India. He might also have begun to believe that he actually was divine, not merely the object of his own propaganda. These glories continued through the entire time of Alexander in Egypt.

On first entering Memphis, Alexander had detailed his lieutenants, Kallias and Petros, to make contact with Barak's metal and leather working shops in Memphis with the purpose of extending Alexander's personal invitation to join him as a principal armorer for his army.

Barak responded to their entreaties that he join them on the march to the

East with gratitude, but also with concern about his family and his business. It would be very difficult to leave this community where he had spent his entire life, had a wonderful family, and a strong business providing needed services to his community.

"I can assure you that Alexander will make these things right for you. At the end of the campaign, you will be returned to your home, your family, and your business here, and they will have prospered within the protections of the army we will leave here, and the contracts for tools, equipment, and services that such an army always needs," Kallias said. "Alexander has also suggested that you and some of your family join him on his trip to the seacoast, and then to the oasis at Siwah in the western desert. Alexander wants to pay his respects to Ammon's oracle, there. Can I assure him of your acceptance of his offer?"

"I have no wish to keep you in suspense. Of course I will accept the invitation to travel with your king, Barak replied. "I would like to bring my oldest son, Nikola, and my daughter Aello with us if that meets with your leader's pleasure. "

Kallias responded. "We serve at the pleasure of our king, but we will advise him of your interests in the trip, and that you have not rejected his offer to merge some part of your metal working shop with our campaign."

"I think you can be more positive about my interest in your campaign than you indicate by your tone of voice. I am very honored by Alexander's offer, and I can see much of benefit for myself and my family. Thank you. I'm sure I will have an answer for you, and for your king by the time we return to Memphis from Siwah. Will I hear from you soon about trip details?" Barak asked of Kallias.

"Yes. One more question, if you don't mind. Why are you bringing your daughter with us on this trip? Does she have interest in travel?"

Barak thought for a moment before answering. "My daughter is very bright, and very curious. In addition to her studies she has been working with me in my shop since she was seven years old. She is very skilled in leatherwork and the making of shields. She has the design skills for flags and pennants. She seems to have a particular interest in the metallurgy of sword making, and in the practice of making tools and weapons out of iron. I cannot imagine going on such a war campaign as you describe, with such important responsibilities to the troops in your army without having access to her knowledge and skill."

"Oh my dear sir. I am delighted with your answer," Kallias responded, "and I am sure my king will be as well. Our concern will be with her comfort and protection, I can assure you, but are you sure that she will be comfortable enough in field conditions over a period of several months… even years?"

"I will discuss all that we have talked about today with my family. You will all have an opportunity to speak with Aello at length on the trip to Siwah. You can judge for yourself about her maturity, and ability to be effective in the field.

"Until I hear from you, then?" Barak asked.

"Yes. Of course, Barak. We will talk to you again as soon as we have the details worked out. Thank you for hearing us out."

=====

After reporting on their first meeting with Barak of a few days ago, Kallias and Petros had been sent by Alexander to inspect the forging workshops and methods of weapons and tool production used by Barak and his work crew.

The men stood in one corner of a crowded workshop in a large building on the Nile River docks near the City of Memphis. Many workmen, sweating heavily, were busy hammering glowing pieces of metal. The heat was unbearable. The men were naked under the leather aprons that all wore to keep the sparks and hot pieces of metal off their bare skin as their hammers hit home.

Barak stood aside as his son, Nikola, hammered a glowing piece of bronze into a knife. Kallias and Petros looked on while Barak reminded them that bronze is a compound of 90 parts copper to 10 parts tin melted together over a very hot fire.

Barak knew the two men were generally familiar with the metal-working process, but he wanted to make sure they had no illusions about the difficulties of creating a portable metal-working system. He continued by noting for the two men: "Such hot fires require several mature trees rendered into charcoal. Charcoal is constantly fed into the fire, and the copper and tin ores are thrown in together in a clay vessel buried in the burning charcoal.

"After a few hours, the molten metal can be poured into molds, and allowed to cool. Once cool we can begin the shaping process. That is what most of these men are doing. It Is not a process very susceptible to the needs of impatient men, especially when they are under attack by a hostile enemy."

Kallias spoke. "It is not particularly well-known, Barak, but one of Alexander's greatest skills as leader of a conquering army is his talent in organizing the materials and supply logistics to support his army while they are on the march. There is no question that he needs the constant support of people highly skilled in making and repairing weapons and armor, but we know these people also need stable sources of materials and stable places to do their work.

"He has seen the quality of your work, and wants to have access to the products of it as frequently as possible while on the march. "Can you talk to me about how you might provide the level of flexible manufacture that my leader needs?"

"That will be a constant challenge to all of us, my friend," Barak responded. "Part of a workable solution will depend on how much transportation support in the form of heavy wagons, horses, and teamsters your king can provide to us when the demand for them during battle will be very high.

"If it becomes necessary to move, it will take us at least a day to have wagons loaded and be under way. It will take at least another day to set up again and fire our kilns at a new location. Add in whatever time it will take to move from the old site to the new. I believe the take down and set up time will fall as we begin to gain experience with the frequency of the demands for new weapons and armor, or the repair of the old ones.

"Of course, here's the other issue that may affect my decision on joining you, or the interest of your King in inviting me. Your army must already have several metal-working and armaments units. I have little interest in working through whatever administrative apparatus you may have for keeping them all in harness, so I hope I will have privileged access to your leader, perhaps through you, whenever I am running into obstacles in the way of providing him with the services he needs. Do I have your understanding on this?"

Kallias answered. "Of course. I will carry that exact message back to Alexander. Now, I am going to have to leave this room. The heat is growing unbearable, and I feel the need to jump into a cold bath, lest I faint dead away."

"We can accommodate you, if you wish. There is an outside pool filled with cool water near where I and my daughter keep our ledgers and do our design work. You and Petros are welcome to use it. Come with me."

=====

As the men approached Barak's office, Barak yelled out to Aello that he was bringing in some guests.

"You are about to meet my daughter, Aello. You recall, Kallias, that having her with me, should I choose to join your army, is not negotiable. She is not only one of my best craftsmen, she is also my right hand in the operation my business, and in our experiments with the making of steel."

When he opened the door, Aello was putting on a light shoulder wrap that would cover her bare breasts. "Oh, I am sorry father. Please excuse my immodesty, gentlemen, it is the heat today."

Kallias seemed unmoved by her discomfort as he smiled and offered greeting, but Petros seemed embarrassed and distraught. He quickly looked away. "I am very sorry, Miss." he said.

Aello smiled in an attempt to put Petros at ease. She rose to shake hands with both men. Barak was glad to see that she was wearing a waist cloth. She was not always so careful when the weather outside combined with the heat in the workshop got these rooms this hot.

"We will use the pool. Thank you for your hospitality. Before we return to our garrison, we would like to discuss two other matters of interest to our King.," Kallias said.

"Of course. Iva, my wife, may have some beer and bread to share. You are welcome to join us."

"Thank you. Petros and I will be very pleased to join you."

====

Later. The two men, looking very much refreshed, rejoined Barak and Aello.

Iva entered the room with a tray of dates, olives, bread and beer. All joined together to share the welcome treats.

Kallias spoke. "Our King has asked me to inquire as to your skills with making steel. Our experience with bronze weapons and armor is extensive, and our troops have learned many important lessons in their manufacture and use. We know that steel is lighter and will hold its edge much longer than bronze. Do you know of these things?"

Barak replied. "I have begun to explore the use of steel, and I have found much to recommend it's use in the field over weapons made of bronze. However, It is somewhat more demanding in the manufacture. The iron ingots must be heated with coal then hammered, cooled, then reheated, hammered, and cooled again. This process must be repeated several times to get rid of its impurities and refine its shape into a useable instrument.

"The process of shaping iron ingots into useful steel is very physically demanding. We are trying to perfect ways to make high quality steel in quantity without killing our steel-makers with overwork. One of these methods involves quenching the steel in a saltwater bath rather than hammering, but we need more time, and more experimentation.

"If you would like I can introduce you to Tabu, who I am training to work with steel. Once you see Tabu and some of his work, I am sure you will understand something of the physical strength I think is required to make steel of weapons grade most efficiently.

"Steel has to be lifted out of the fire, and held, then heavy hammers have to be swung steadily and quickly to shape the piece. The faster that Tabu, or any other smith can make that happen, the sooner we can produce a steel weapon or piece of equipment with sufficient quality for heavy use in the field.

There is another factor, perhaps most important of all. Iron ore is usually plentiful, but not always easy to find, or to remove from the earth, or to move from the site of a mine to the site of a smelting furnace, so it can be turned into pig iron. I have no knowledge of Alexander's proposed route, not do I have any idea of the kinds or qualities of ores we might find.

"What I do know is that while we are advancing, Alexander must constantly judge our need for raw copper, tin, and iron ore with which to meet his demands for weapons. I am certain he will maintain reconnaissance looking for the enemy, and for routes through difficult areas. This reconnaissance must also include a fairly rigorous search for the ore needed to make weapons. This has to be an essential factor in all of his battle planning and campaign logistics. If he already has such a unit that he has confidence in, then I need say no more about it. If not, please advise me of what he thinks he may need, and I will prepare a proposal."

Kallias replied. "I will pass your concerns to Alexander."

Kallias continued. "There is no need to meet Tabu now. If he is as good at steel making as you say, I'm sure you will want him to go with you on our campaign. For now perhaps I can suggest that your man Tabu craft a steel sword suitable for inspection by a warrior King. Bring Tabu and the sword along on our trip to the Mediterranean coast, and to Ammon's Oracle at Siwah."

"It will be my pleasure to so instruct Tabu." Barak replied.

Kallias continued. "We plan to depart to the north by ship within the week. Alexander has a great interest in building Greek shipping interests up to the point where we can displace the Phoenicians. For that he will need a major seaport on the Mediterranean Sea, and a strong navy to secure the commerce, that we know will evolve once the port is in place.

"We will be spending time along the coastal area looking for a site for the development of a major port. In addition to his many talents for wartime tactics and strategy, Alexander also has many skills in city planning and architecture. We will have an opportunity to see all these skills played out in full as we work with him and think about a whole new city and seaport." Aello looked at Kallias and Petros and smiled at the possibility of having so many things to learn. Petros' look lingered for a moment. Then he smiled at Aello and looked back toward Kallias.

"You and your party should consider yourselves to be very fortunate." Kallias asserted.

Kallias continued. "Once Alexander is satisfied with his planning work for the new city of Alexandria, we will leave most of the Army to return to Memphis. We will move overland to the west. We think the trip to Siwah might take as long as 10 days. We will be a fairly small group, though Alexander will have some of his best men ahead, behind, and among us.

"We could be in the field for about two months. Can you make your preparations for joining us, and also arrange your affairs in Memphis for a possibly long absence accordingly? When my King asks for your statement on this, he will be very alert to any ambiguity, or lack of confidence."

"Yes. We can be prepared enough to sound very confident of ourselves." Barak replied.

"Good. My King, Petros and I look forward to many conversations with you and your workmen…"

"…and my daughter, Aello. I want to make sure we are clear on that." Barak asserted.

"Of course, we are. My King has already expressed his willingness to include whoever you think must go with us, and certainly the King understands that your daughter is very important to you in your work. Actually, it was he who took particular interest in her knowledge of making steel. She will be welcome. We do know how important it is that we have a good working relationship with you and all your craftsmen.

Kallias bent closer to Barak and spoke just above a whisper. "Our campaign will be hot, difficult, and dangerous. We have enjoyed great success in our battles so far, but I feel the success of our future campaigns may well depend on our adaptability to our surroundings as we get farther from our home territories, and the strength and flexibility of our supply lines. Our success will depend absolutely on the level of commitment to our cause by you, by each of your craftsmen, by your daughter, and by all who are both loyal and technically skilled in the Strategy, Tactics, war making details, and day-to-day minutiae of war-making.

"All will be challenged at their weakest points by our very capable enemies with their large, and very well-equipped armies.

"Alexander also believes this. You must trust my word."

====

In January, 331 BCE, Alexander's party of guests and his contingent of hand-picked Macedonian soldiers departed Memphis on two quinqueremes of the Greek navy. These ships with their crews of perhaps 200 men working five rows of oars in shifts were the largest and fastest of Alexander's Navy. Among the other triremes and quadriremes that made up Alexander's expeditionary fleet the quinqueremes were the flagships.

The Egyptians had given Alexander a royal and honored welcome on his entry into Egypt in October. In naming him Pharaoh in November, the Egyptians took particular note of the rapid rout of the Persian forces under Darius III that soon followed. He felt no need to beef up security on the way down the western-most channel from Memphis through the Nile Delta to the small port of Rhakotis.

Alexander had a vision of a great new city and port on the coast of the Mediterranean. The existing port of Rhakotis was adequate for the level of commerce and shipping now working in the eastern Med, but Alexander had much bigger plans for the north end of the huge, grain-rich agricultural region known as the Nile River. A talented city planner and architect in his own right, he had brought with him a delegation of planners, architects and engineers familiar with building large structures on soft ground.

Among Alexander's other personal guests on the ship were Barak, the Greek Jew who owned a large shop of armorers and iron-mongers in Memphis. Also with Barak were his son Nikola, and his daughter, Aello, both highly skilled and experienced in the arts and crafts of war in their father's shop.

Barak's Nubian chief sword maker in bronze and, in recent months, in iron and steel, Tabu, was also with Barak's party. Tabu had completed his work on the steel sword that Kallias had suggested back in Tabu's workspace in Memphis. The sword had only to be presented to Alexander for inspection as also suggested by Kallias.

Barak knew that of all the weapons available to him, Alexander preferred the sword. His current personal sword, called a kopis, had a curved blade made of bronze; reputed to be excellent for lopping off limbs and heads. Barak knew that bronze - a composite made by smelting copper with tin - was relatively heavy and brittle. It often broke in battle.

Worse, lots of tin was required for the making of a bronze sword, and tin was in short supply. In the field on a long march far from supply lines, the scarcity of tin for weapons could be a fatal problem. Alexander knew that he needed something better as the primary personal weapon for his troops, but so far, none of his armorers had anything better to offer. The curiosity of Alexander became Barak's opportunity, and Tabu's challenge.

Tabu's presentation sword was simple. He and Barak believed that Alexander was sophisticated enough in his choice of weapon that he did not need to be distracted by decoration, or by extra, unnecessary weight. As Tabu worked the steel by constant heating, hammering, and cooling, it began to take on the shape of a sword. As he got closer to putting an edge on his work, he began comparing his work to a bronze kopis that Kallias had loaned them for that purpose.

Tabu's steel sword was straight, and the blade was narrow, unlike the curved blade of the kopis that widened toward the tip. Tabu had been a warrior in the tribal wars of his Nubian homeland at the headwaters of the Nile. He had experience with weaponry, and with the death and disfigurement that a warrior could bring with a good weapon, but he had never had a weapon that felt to be such a part and extension of his body; that had such an instinctive flow as he swung it toward an imagined foe. He was very proud of his work.

Aello wanted to learn all about making steel. She helped Tabu with his work with the steel in any way that she could. At the end of the process, Tabu asked her advice on some kind of a message to be engraved on the body of the sword that would be pleasing to Alexander.

Aello conferred with one of her most trusted fellow scribes about languages, phrases, and writing styles that would be appropriate in a gift to the newly named Pharaoh of Egypt. The scribe suggested a phrase and a style for a script to be written in Greek. He also suggested a Greek engraver who could do the work on the finished steel. Tabu was very pleased with the results of Aello's recruitment work.

"Let me know when the work on the sword is finished. The engraver, Pelagios, will need a couple of days. We will need to have the sword ready for presentation by the time we leave for Rhakotis in January." Aello said to Tabu.

Tabu returned to the hammering on the presentation sword with renewed enthusiasm. He loved Aello in his way, but his way was one of homosexuality. Since Aello had barely begun to walk, he often found himself acting as a father figure and protector for the inquisitive and beautiful young daughter of Barak, his employer. Barak in his turn welcomed Tabu's protection. His daughter's curiosity combined with her beauty and intelligence occasionally led her into difficult situations in the marketplace. Tabu, with his sheer size and power had always been able to resolve them, though not always without some violence to the perpetrator, or perpetrators, of an incident against Aello.

Again because of her inquisitiveness and her great affection for Tabu, she followed him into the workshop at every opportunity. Over time, with Tabu's guidance, Aello became familiar with all of the armor and weapons produced by the workshop, and, more importantly, with all the production methods and personnel who made production possible.

When Barak recognized the talent and experience that Aello was bringing to the workshop under the mentorship, guidance, and love of Tabu, he brought her into the very center of the business. These are the arrangements and finances that allowed the workshop to keep doing high quality work in high demand over a period of years. By the time of Alexander's trip down the Nile to Rhakotis, Barak had total confidence in Aello's ability to run any aspect of the business of the workshop.

=====

The trip down the western branch of the Nile to Rhakotis on the flag ship of the Pharaoh of Egypt was uneventful. The trip took less than a day. On shipboard Alexander was engaged in discussions with his city planners and architects and was not available to his other guests. Kallias also had duties in attendance to those invited to the meeting and was also unavailable to other guests. Petros had the job of entertaining those not at the meeting. That included Aello; a prospect that Petros felt both excited and nervous about.

Aello, understanding his shyness, did all she could to put him at ease. She brought him into the planning for the presentation of the sword to the Pharaoh. Tabu had burnished the sword to a dull bluish glow that brought the sword to a powerful life of its' own. The engraving seemed to give the living sword a voice. Petros told Tabu: "I have never seen, or felt, or held a sword that comes to life in your hands and speaks with such authority. Alexander will be very happy with it. He will be proud to have received it."

=====

Alexander and his planners spent two days walking around Lake Mareotis, and the area near the port of Rhakotis. Alexander and his party had found the natural harbor that he was looking for. He began to look for sites for the permanent structures that he envisioned. These included a government building for the administration of the great Macedonian city, and a temple for the worship of the gods of Greece and Egypt. All of these structures and all of the necessary streets were laid out by Alexander and his party.

Once Alexander found sites sufficiently high above sea level to assure that the foundations of the large buildings would have sufficient bedrock below the surface to insulate the buildings from the rise and fall of the water table, he began to work with his Chief Architect Dinocrates and their land surveyors to pick and map sites for individual structures, and to then layout their foundations on the ground. Alexander lacked materials for physical markings, but he made due with grains of barely from the rations of the ship's crews.

On stepping onto the edge of a spit of land lapped by the waves of the Mediterranean Sea, called the Pharos, Alexander's fertile imagination conjured for him a vision of a great tower. The tower would be lit at night with a huge light at the top to warn ships in the area of land; to help guide those ships into the safe harbor and to mark the location of the greatest city ever to be founded by Alexander and named after him; the greatest founder of cities where none had ever before existed.

At Kallias' urging, Alexander had been persuaded to allow Aello to accompany the city planning group. Beyond being introduced to each other, Alexander and Aello did not speak further, but Aello knew that her face-to-face meeting with the new Pharaoh had been the opportunity of a lifetime. She spent as much time as she possibly could trying to comprehend all the things that were being discussed among these skilled and dedicated men.

There were so many new ideas, and she had so many questions, but she knew that any interruptions to their deliberations would ruin the welcome that the good man, Kallias, had prepared for her. She could barely wait to tell her father, Barak, her brother Nicola, and Tabu of course of all she had seen on this wonderful day. She thought of the shy Petros. She wished he could be here with her.

====

Eventually it became time for Alexander to leave his newly named city of Alexandria to move on toward Siwah, far to the west. He discharged several of his military staff and their units. He ordered them to return to the flag ship, and then back upriver to Memphis to prepare for the campaign to continue pursuit of the escaped Darius III and his army of Persian regulars, now thought to be to the east in Mesopotamia.

Kallias took Barak aside. "Alexander has asked me to discuss with you the trip to the oasis of Siwah in the west. This will be a long and very difficult journey over many hot days in the desert. There will be great risk. The desert is trackless. Our only hope will be the experience and skill of our guides. If we lose our way, we may die. There will not be much to learn from the trip. Alexander has enjoyed your company on this trip and will hope to become better acquainted with all of you, but he has told me that you must choose to come with him. Those of us who are trusted to carry these kinds of messages are admonished to convey his wishes to you without any sense of guilt if you choose, instead, to return to Memphis with the other members of the military staff."

"Can you tell me anything more of the purpose of the trip?" Barak asked. "I know my daughter, Aello, has been delighted with everything she has seen and experienced on this trip, and she would feel very disappointed with herself if she leaves by her own decision no matter what the reason. For myself, I can foresee conditions on the campaign to the east that will be far more dangerous as we move farther and farther from the safety and supply of our home territories.

Barak continued. "I believe that I, Aello, and Tabu all need to accompany the Pharaoh on his trip to Siwah as observers. My only concern would have to do with something the Pharaoh must do, that we must not see at risk of our lives, or that we must participate in that will so violate our Jewish faith that we will feel the need to turn away from our most deeply held beliefs.

"Otherwise, I feel we must go with you. We will see the experience on the rugged trail ahead as a final determinate of our agreement to go with you on your campaign to the east. I speak for my daughter, for Nikola, for Tabu, and for myself. Please convey my thoughts to your King, and please express our warmest appreciation for his courtesy and support for us on this trip. We will be forever grateful."

Kallias had been looking at the ground at his feet while Barak spoke. After a moment he spoke. "I had not known of your deep faith, nor would it have changed anything if either I, or Petros, or our King had known of it. The House of Alexander's father, Phillip II of Macedon assigned the philosopher Aristotle as Alexander's tutor.

"Aristotle has always found great value in the idea of a creator God, as do the Jews; as does Alexander, though his advocacy of particular faiths is more pragmatic. Alexander has also had the opportunity to speak to many Jewish leaders in Jerusalem, and to receive their blessing when he and the army passed through there. He believes that the loyalty and support of the Jews continues to be his guarantee of future victories.

"We greatly value the work of the Jewish people in their many services to us. I believe the presentation of Tabu's sword to Alexander will re-confirm all in that long held opinion. I will personally see to the arrangements for that event before the army departs for the return to Memphis."

"Thank you for those kind and encouraging words." Barak said.

Kallias continued. "Let me describe the purpose of our trip to Siwah. If you and your family and Tabu will then re-confirm your desire to go with us, I will convey that welcome news to our King."

"Alexander will be following in the tradition of other Macedonian Kings before him when given great honor, as the Egyptians have honored him by naming him Pharaoh. He will confer with the Oracle designated by the great god Ammon. That conference will be the sole purpose of the trip. Though I am assured that Siwah is a beautiful place, those of us who accompany the Pharaoh on the long trip across the desert may have little else to be grateful for.

"No one will be allowed to accompany the Pharaoh as he confers with the Oracle. You are all welcome to accompany us, but we cannot, in good faith, guarantee the success of this mission, nor can we guarantee the safe return of you and your loved ones to your homes and loved ones in Memphis."

Barak spoke: "Please convey our grateful thanks to the Pharaoh for his kind invitation. We will be very pleased to accompany him to Siwah. If you wish to see the sword which has been hammered by Tabu, come with me now."

Kallias replied. "I am afraid that is not possible. Alexander will want the sword presented directly from the hand of its' maker, Tabu. For a member of his staff to lay eyes on it before him, he may see it as having been used before he has had the opportunity. Believe me, I will take the greatest joy in seeing the sword at its presentation."

=====

"How was your time with the great man during his walkabout in the new city of Alexandria, my daughter?" asked Barak."

"He is a beautiful man, father. He is not a tall man. Tabu would tower over him, but he walks as though he contains all the fires of your many forges, all throwing sparks with the sounds of all the hammers all at once." Aello replied.

"That is high praise. Will you be able to work effectively with him? It sounds like your first view of him caused you to swoon." Barak continued.

"No father. I think I know exactly what I am feeling, though I admit I am less sure of myself in working with him. I know that anyone who gets close to Alexander can easily be consumed by him. That person could end up with no more value to himself or herself, or to others than the slag hammered off the hot ore as it has been taken from the forge."

"On the other hand, the intensity and heat of his personality is not all there is to the man himself. He is an intelligent and capable man of great vision. If I learned nothing else during the time we spent over two days walking around the hot sands of Lake Mareotis, it is the fullness of his vision of his new city by the sea. On seeing the secure harbor and the safety of the approach from the sea through natural breakwaters, from that moment forward he literally carried an image of the docks and the whole city in his head. When he paused to consult this mental image, he only did so to instruct the architects and surveyors where each street center-line and foundation corner must be placed.

"These are not the things about a man that should worry a father concerned about his daughter, however, father. Those things have to do with Alexander's humanity, and the warmth he shows toward the men and women around him.

"I can assure that you should worry, father, because I worry at these behaviors, and I did worry about my ability to think clearly when I stood close enough to him to hear his voice as he directed someone to move a pointer, or stadia measurement device a little forward, or a little backward.

"I have one example for you, father. One of the architects was standing on the rocks overlooking the waves splashing onto the shore below while writing notes in his book. I'm sure it was an accident, but he suddenly dropped the book into the water several body lengths below, too far down to retrieve. The architect feared the worst. He sat down and put his head in his arms.

"I was nearby and saw a look of pure fury briefly fly cross Alexander's eyes. I shook with fear for the note keeper. Even so, he walked over to the man and put his arm around his shoulders to calm him down and get him back to work with a new book. After a few minutes, both stood up, the architect began making notes once again, and Alexander walked back toward us. As he passed by me, he winked. My knees got weak, father, and I almost passed out."

Barak commented. "Perhaps he calmed down only because he knew you were watching him."

"I am sure that is a possibility. He is the leader of one of the very largest armies on the earth, and there are many stories of the terrible violence, beheadings, and disembowelments he has done to his enemies in the opposing armies, even to his own soldiers who have become disloyal. Those stories do not reflect the character of a gentle or kindly man. I'm sure he will not be so forgiving with those he engages in battle, even those on his own staff if he knows he must make an example of them in order to avoid very ugly consequences."

Barak spoke again. "I am glad you told me these things, my daughter. I think our trip to Siwah with the Pharaoh will be important to our final decision on who will go east with his campaign. For now let us prepare for the presentation of Tabu's sword. After that we must finish preparing for the trip to Siwah."

Aello stepped forward to hug her father. "Thank you, father, for taking the time to listen to me. I wish I could capture your words as you speak them. I would paste them to my body so I would always have them where I can always see them."

=====

Two days later, on the night before the Pharaoh's departure to Siwah, Kallias had finally managed to get everyone together for the presentation. It would be a fairly small, fairly quiet gathering among some trees in a shadowed copse next to a stream flowing into the western most channel of the Nile Delta. The presentation of a new sword hammered in ways specific to the character of a Pharaoh required dignity.

There would be no speeches; no formal introductions. The Pharaoh walked up to a small stage under a tent and took his seat. He was dressed casually, though he did wear a bronze cuirass, and his belt held a short sword. His sun-darkened skin contrasted sharply with his curly, sun-blond hair. He is a handsome man, very sure of himself, even when seated, though he sits as though he may need to jump into action at the slightest provocation. He is very young; only 24 years old. Kallias and Petros stepped behind the Pharaoh, and to his right. A four-man detail of soldiers in full armor and battle weaponry spread themselves along the rear of the audience.

The audience stood in front of the short platform. Barak, Nikola, and Aello in clean white robes and jewelry and headdress suitable for an audience with the Pharaoh. Tabu, a black man dressed in the most formal workman's robe available, stood at the center of the gathering, his arms extended in front of him holding something narrow and long as an arm, a piece of white cloth draped loosely over it. Various other members of the audience were specially invited by the Pharaoh, or by Kallias.

Kallias motioned to Tabu that he should step forward, approach the Pharaoh, and kneel with his head bowed down, with the sword extended toward the Pharaoh for presentation. Tabu, almost overwhelmed with the dignity of the ceremony and his place in it, did as Kallias indicated.

Alexander rose from his seat. "Please stand Tabu. I am told you are a maker of fine steel swords. To a soldier like myself, such a man deserves to stand eye to eye with me, even though your height puts your eyes a ways above mine. What do you have for me?"

Tabu, afraid to open his mouth in response, cradled the sword with his left arm, and took the cloth away from the sheathed sword with his right. He dropped the cloth, but kept his head bowed while he extended the sword, still cradled in his left arm, to the Pharaoh. Alexander took the sheathed sword, then clutched the hilt with his right hand, and pulled the sword out of its sheath. Tabu dropped his arms to his side. His eyes remained downcast.

Alexander turned to his left away from Kallias and Petros, swinging and twirling the sword lightly with his wrist as he stepped off the platform, and moved away from the gathering. Suddenly, he swung the sword hard, first left, then right. He walked to a tree as big around as his forearm. He swung the edge of the sword, hard, at the trunk at eye level. The top fell. Before anybody could react, he continued in the same motion by swinging at the tree again backhanded. Another piece of the tree fell.

He tossed the sword in the air and caught it with his left hand in one sweeping motion aimed at another tree, of the same size. The top of the other tree fell. Again he swung in a single motion backhanded, and another piece of the second tree fell.

Alexander turned and called out to Kallias. "Kallias!" he yelled. "Borrow some armor and a kopis from one of the troops there and come face me. Be ready to fight"

Kallias did as he was bade to do by his King. He is an excellent swordsman, and Alexander often used him as an opponent. Both men understand that these kinds of intense but not fatal sword fights were how they kept themselves in fighting trim. They were also important in testing these two personal swords; the bronze, wide-blade kopis held by Kallias, a personal killing sword, a standard design distributed among some of the infantry and all of the officers of the Greek army. And this new sword hammered, personally, into sharpened steel out of hot pig iron by Tabu already put by him into the hands of the Pharaoh.

Alexander walked back toward the group, again lightly swinging and twirling the blued steel while he looked along the length of it. He read the engraving aloud: "Plato is dear to me, but dearer still is truth."

As he returned the sword to its' sheath, he said "A fine statement by my teaching master, Aristotle; a fine statement for a fine weapon. Thank you, Tabu. I hope you and your master will agree to join us on our campaign to the east."

As Alexander turned back to face the approaching Kallias, now armed and armored, Tabu spoke. "Thank you, my Pharaoh. It is my great honor to serve you," Tabu said.

Alexander's first thrust to the chest was easily parried by Kallias who then moved to block Alexander's next killing thrust low toward Kallias vital organs. The two men were closely matched, and well aware of the many practiced and usually successful thrusts and parries against enemies that they themselves had to protect themselves against. After a few minutes of intense swordplay, Alexander suddenly struck, hard, at Kallias' kopis. The bronze blade of the kopis broke under the power of Alexander's stroke with the steel blade. The broken blade fell to the ground and lay there.

Alexander's theatrical nature broke through briefly as he twirled his sword with his wrist, then thrust his sword against the now undefendable neck of Kallias, stopping within a finger's width of raw flesh, and laying the flat of the blade against Kallias' skin.

Alexander bent to pick up the broken blade of the kopis. He noted that there were several dents in the cutting edges, and, ultimately, a break when the steel blade hit the two opposite sides of the bronze blade.

"An excellent test, my brother and friend." Alexander said. "This tells me all I need to know about this new metal blade that our new friend Tabu has crafted for us. "

Alexander departed quickly with his four body guards, and Kallias.

The crowd also disbursed. Before leaving, Kallias held Tabu aside with a brief comment and smile. Tabu seemed embarrassed, but very pleased with

Kallias' attention. Tabu joined Barak, Nikola, and Aello. They walked back to their quarters feeling great pride in the fine piece of work Tabu had done.

The next day they arose early for the 10 day trip along the coast of the Mediterranean Sea, then south to Siwah. The trip began and continued routinely along the seacoast. But the routine was not to last as they turned south away from the coast, and across the hot, dry desert.

====

Desert caravans are a wonder to behold. Though small, Alexander's caravan on this day was no exception to this visual splendor. The desert guides were nomadic Berber tribesmen. They wore long colorful robes wrapped around their bodies and multi-colored cloths around their turbans to keep out the blowing sand. They could move across the desert without guideposts ahead, and without leaving tracks behind. As was typical with Alexander on a long march, he joined the Berber men walking more often than he rode. Not to be outdone, the men with Alexander, including Barak and Nikola, also walked. Aello walked.

Those who rode were mostly Berber women, their small children, and some very old men. Pack animals, mostly burros, loaded with gear, provisions, and weapons, sometimes with children, made up the bulk of the caravan. Caravans are often noisy and colorful affairs. The camels, the burros and any other livestock had colorful pom-poms and woven, multicolored leads attached to their halters and pack saddles. The Berber men often broke into their native songs, and Alexander, enjoying the spectacle, commanded his men to join in with Greek drinking and marching songs.

Four days south of the coast, a sand storm began to form in the west, and move toward them. Typically, it moved, in the distance, in high, roiling brown clouds. The Berber guides, recognizing the danger, knew they would have to stop and secure the material in the caravan against the wind and blowing sand. They would put up tents, heavily anchored by long pegs hammered deep in the sand, for the people. The animals would all need to be hobbled. The guides asked Alexander, through his multi-lingual interpreter, Petros, if he wanted a separate tent. He declined the offer.

"I'll stay in the tent with my soldiers and our guests," he said.

The wind and wind-borne sand blew strongly for more than a day. People in the tents could hear the sand blowing over the tents like an abrasive cloth. Some feared the tent material would go weak from the abrasion and tear. Fortunately, the tents did not tear.

Eventually the storm abated. As the people in the caravan began to crawl out from their tents, the guides conferred with Alexander through Petros. They pointed out that the landscape had considerably changed, and they may have lost some of the landmarks they rely on. Without landmarks they could not guarantee that they could find the oasis at Siwah. Worse, the guides feared that the storm would have shifted the sands enough to cover up two water holes that they normally expected to find along their regular route. Without those water holes, the chances of getting to Siwah, even if they knew the most direct way, were slim.

Alexander's response was to order his men to take a full stock of their remaining provisions, particularly water. Alexander told the guides that he had two talented scouts with him and they could find a trail in any kind of terrain. The guides pointed out that the sands were blown into a patchwork of hills and depressions; none of the tribesmen had seen before. Any scouting party will quickly lose any contact with the main party. The rolling dunes and lack of vegetation to make smoke signals or platforms means that their ability to find the main party again will be poor.

Alexander pondered these difficulties. Alexander's staff reported that provisions of food were adequate for another 10 days, but supplies of water were poor, perhaps only another five days under careful rationing. The camels would be fine, but the burros and other pack animals would not likely survive if they could not find the first of the water holes the Berber guides are relying on.

On hearing of these new challenges, Alexander knew that everything about his life and his vision of the future Greek world would be at stake on his next few decisions. But his youth, and his belief in his likely divinity gave him courage.

He called the Berber guides back to his side. "Can you continue on toward Siwah, working with my scouts and using your best judgement about direction; and can you find the first of the water holes you think lie ahead of us?" He asked.

The lead guide spoke through Petros. "Yes, but we are concerned about the welfare of our families. We must decide between trying to return to the coast, or continue with you to Siwah. We feel the return to the coast is our safest choice. Of course, we must also consider the blessings the Egyptian Pharaoh can bring to our people if we stay with you, and we are successful in bringing you and your party to Siwah."

Alexander spoke: "Prove to me that you know the direction we must travel to the next water hole. Do it now when the sun is high in the sky and offers no shadow clues."

Without hesitation the guide turned slightly to his left and pointed straight ahead.

Alexander made the lead guide an offer that might justify a decision in his favor, then dismissed the guides so they could confer. After a period of time, the guide indicated to Alexander that they would accept his offer and continue with his party to Siwah.

Alexander issued instructions to his staff to begin loading the animals and preparing to continue the trip to Siwah. He then moved away to a dune some distance away in the direction indicated by the lead guide. He sat by himself. He knew he would not be disturbed until the entire party with their guides were ready to move sometime after the hottest part of the day. If there is moonlight they might continue to travel until there is no more light to see by.

Alexander had some understanding of the positions of the stars as gained from the careful teaching of his tutor, Aristotle. Perhaps they would show him a path he could follow. If not, the travelers would then sleep until the first light of dawn, travel further, then sleep under cover if too hot to travel until the day begins to cool.

Alexander took note of a flock of about a dozen ravens flying in circles far above the dunes ahead of him. They were in the direction the guide had indicated to the first watering hole. He interpreted this as an omen of good luck.

=====

Four days later, just before the dawn began to lighten the sky, the first water hole remained undiscovered. Alexander, groggy and thirsty, his youthful, patchy beard unshaved, rose to see the direction of the rising sun. He stumbled outside. He heard more than saw the ravens circling and gabbling nearby, just as they had since beginning this part of their journey. Aello, who had seemed to suffer the least from the difficult journey, joined him. She had wrapped her body in one of the colorful robes loaned to her by one of the Berber women.

"Good morning, my Pharaoh," she said. "I have been watching the ravens. One of them flies outside the main group. It looks at me from time to time." She paused, in thought. "I think I can talk to it."

Alexander was not ignorant of the possibilities of such things. His mother, Olympias, had told him he was destined for greatness, even divinity. After the Egyptians made him Pharaoh, he had begun to think of himself as God-like. Indeed, his near Godliness was the very reason he wanted to speak with the Oracle of Ammon in her temple at Siwah, just as the greatest of the Greek Gods Zeus had done so long ago. He resolved to complete this part of the journey on the way toward his final destiny.

"If the raven speaks you must listen and learn. Come back to me when you are sure you know his message.

"Thank you, my Pharaoh." Aello rose and walked in the direction of the circling ravens. When she got near them she sat down at the top of a high dune with her head bowed, her robe pulled close in the cool air. Shortly, she felt the presence of another being. The raven that had been flying outside the group was now a few paces away from her, looking at her, making soft vocalizations. After a pause, the raven turned away, then, extended its' wings and rose to fly away. She noticed a single white feather on his right wing. As he ascended into the sky, he turned again to look at her, then rejoined the circling ravens.

Now back together, the ravens flew in a single direction then descended down among the dunes almost out of sight against the dark sky a few hundred paces away. Aello looked at the sky above the descending ravens. She saw four stars at the end points of a small cross just above the horizon. A few more moments and the stars would fade in the gathering light of dawn. She returned to Alexander's side.

She pointed toward the four stars, down and a little to the left of the Belt of Orion. "We must go in the direction of those four stars. We will find water there."

"Those four stars are sometimes called 'The Southern Cross' by my naval officers. That is a good omen. I will direct my men, and the guides to load immediately and begin to move toward those stars." Alexander responded. "You must remain here and search the horizon for landmarks under those stars that we can follow after the stars fade away in the dawn. I will detail two of my men to stay a distance away from us on either side, and on top of the dunes to stay on the lookout for signs of water. You must concentrate on the location of the ravens. If they move in a new direction, you must find me, and tell me immediately."

"Yes, my Pharaoh."

Later that day, the travelers discovered the water, and that it was plentiful at the watering hole. The guides said this appeared to be a new watering hole on the route to Siwah. They were pleased to see the still gathering pond.

Alexander ordered a day to recover and regroup.

====

It is almost 2 years after Alexander's ordeal in the desert, and his visit to the Oracle at Siwah. Alexander never did speak of his conversation with the Oracle;...not with anybody. He caught up with the fleeing Darius III, finally, in a tough battle at the Caspian Gates 40 miles south of the ancient city of Teheran. Darius was already dead, killed by his own people, or by his own hand. It was not clear how it had happened. Alexander honored him as a valiant warrior and buried him with full military honors.

With the death of Darius, the Macedonians in Alexander's army wanted to go home. In their minds the war was over. In the mind of Alexander, and in the minds of his younger soldiers loyal to him, however, Darius was gone and his army was in total disarray. The way was now clear to roll up the rest of Persia and take the Greek army all the way to India. In that way the Greeks would have an empire even greater than the Persians.

Alexander was ready to move on to India, but first he would have to secure Afghanistan – a place of high mountains, and regional tribes lead by strongman warriors. In order to establish the civilizing influence of the Greek city-states among these warring tribes, he would need to keep the Macedonians with him, and willingly under arms.

Since leaving Egypt, Barak and his metal and leather working company has been closely attached to Alexander's army. He has continued to gain the favor of Alexander and his closest military advisors. With daily operations under the supervision of Barak's son, Nikola, Barak has been free to work closely with Alexander's operations planning staff to find the sources of ore and other materials necessary for the manufacture and maintenance of weapons, armor, and other battle gear.

Alexander has also found great favor in Barak's daughter, Aello, since their time together, along with some of Alexander's elite military guard, as part of a guided caravan traveling from the Mediterranean coast to Siwah. Aello's observations and careful listening to the gabble among a flock of passing ravens helped Alexander rediscover a path to Siwah that had been obliterated by a devastating sandstorm.

Increasingly, Alexander grew more firm in his belief that he is the son of Zeus, the most important of all the Greek Gods. Though he did not talk of his visit to the Oracle of Ammon at Siwah, whatever had been discussed there did not discourage him in his belief in his own Godliness. This belief had been growing since the Egyptians had named him Pharaoh 4 years before; had intensified with the successful battles at Gaugamela in Mesopotamia and at the Caspian Gates near Teheran that marked the final defeat of the Persian army.

Alexander continued to think of his interaction with Aello, and her interaction with a flock of Ravens when they and their caravan were lost in the desert as a further indication of his godliness. This motivated his favor toward her. His favor did not extend to a sexual interest in her, however. Alexander had an ambiguous view of sex and used it more as a way of extending power – with either men or women - than as a source of pleasure. Also, the sexuality among the Greek gods was a complicated affair, and Alexander probably did not want to temp his fate among the Greek Gods by having a sexual liaison with a woman, even though she is as beautiful and sexually attractive as Aello, who also seemed to speak to Ravens, and hear their responses.

For now, it seemed to Alexander – when he thought of it at all – that Aello seemed to be happy in her role as liaison between Alexander's operations planning staff, and the important work of Barak and his workshop of weapons makers. She had also found pleasure in her relationship with Petros.

As Alexander's army grew increasingly diverse in the ethnicity and language of the local armies drawn under Alexander's banner, Petros' work as a multilingual interpreter became increasingly important. Whenever he went with Alexander, or his staff, on a diplomatic mission, Petros asked, usually through his friend and colleague, Kallias, to have Aello accompany them. Alexander would want her to evaluate weaponry needs, but actually he knew that her presence would be an expression of their deepening bonds of friendship and love.

Unfortunately, the many questions of who among all of Alexander's senior staff would accompany his army on their long march into the interior of Afghanistan and across the Hindu Kush mountains threatened to create many divisions among long-time colleagues, friends and lovers.

As a military commander of an army in the process of conquering much of the known world, Alexander knew that such considerations could not be given much weight. As the son of Zeus, he also knew that all errors of judgement would eventually be forgiven.

PANEL 4 - FAITH

The young, impetuous Gila and her older sister Ziva meet Jesus of Nazareth on a quiet street corner near the lower market during Passover in Jerusalem. Later, both girls and their Jewish families are taken into domestic slavery in the family home of a Roman Centurion.

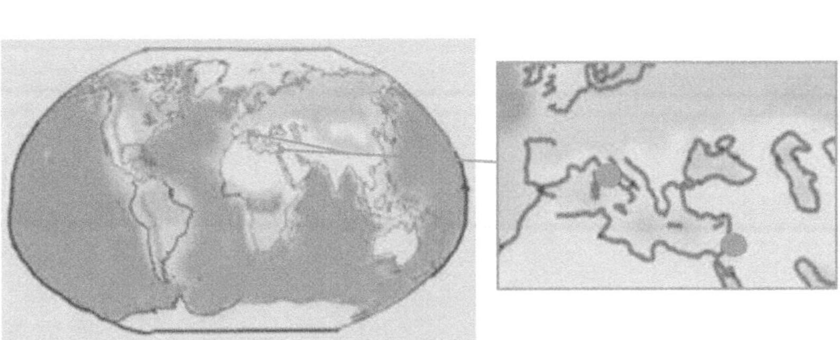

ROME AND JERUSALEM. SPRING 20 CE.

PROTAGONIST

Gila - 10 years old. The year is 20 CE. The first day of Passover in that year is Friday, April 26, in the modern calendar. Ziva - Gila's older sister.

Avraham - Gila's Father and Orli -Gila's Mother.

Lucius – Son of Quintus who is a few years older than Ziva and Avi. Gila's younger brother.

Bathsheva and Aharon - Friends of Gila's family who live in the lower quarter of Jerusalem, also known as the Jewish quarter.

Jesus (b. 4 BCE– d. 30/33 CE) born in Nazareth and known outside the Galilee as the Nazarene, a charismatic healer, exorcist, and itinerant prophet who works as a carpenter and day laborer on construction projects in nearby cities, and among the fishermen around the Sea of Gallilee.

ANTAGONIST(S)

Quintus Caecillius Bassus – The Roman Army Centurion Primus Ordinis of the second cohort of the 10th Roman Legion based in Syria., Cornelia – Wife of

Quintus. Decanus (sergeant) Kanutus - a soldier in Quintus Cohort. Eventually at

Quintus orders, Kanutus takes Gila and her family into custody. Axius and Marcus,

Legionnaires with Kanutus

Theodorus - A Tribune of the 10th Legion who reports to the Legate in Caesarea Maritimus who leads the Legion, and, informally, to Tiberias, the Emperor of Rome, and a fellow member with Theodorus of the high-born classes there.

Kaius - The Legate, a senator appointed by the Emperor. Kaius commands the 10th Roman Legion in the Province of Syria, headquartered at Caesarea Maritimus.

Pontius Pilate – (b. ? - d. 38 CE) Governor of Judea from 26 CE to 36 CE).

LESSER CHARACTERS

Kema – An older Egyptian slave woman; Loyal housekeeper to Quintus; put in charge of Gila's up-bringing and training. Kema is with Quintus in Jerusalem, looking after his household.

Shani - A young African slave girl in the house of Quintus under the supervision of Quintus' wife Cornelia.

Anyim - A young African slave boy in the house of Quintus, basically a handyman under Quintus direct supervision.

Timaeus - Centurion Primus Pilatus of the first cohort of the 3rd Roman Legion based in Gallica.

MINOR CHARACTERS

Valerius Gratus - Governor of Judea until replaced by Pontius Pilate in 26 CE.

Caiaphas - Last High Priest of the Temple in Jerusalem appointed by Gratus, after a series of hiring and firing of priests that proved to be troublesome during Gratus rule.

Ephesius - A tribune on the staff of the Prefect Pontius Pilate.

After the assassination of Julius Caesar by the members of the Roman Senate in 44 BCE, a young nephew, Octavian, then 19, had been named by Caesar, in his will, to succeed him as Emperor. Octavian renamed himself Augustus, abandoned the ideals of the Republic, and took over the running of the expanding Roman Empire. Despite the bloodshed that attended his ascent to power, and his trashing of Rome's Republican Senate, history does not treat Augustus poorly. He kept the peace during his 40-year reign, under a beneficent martial law, and he expanded the empire to include Egypt, northern Spain, and parts of Central Europe.

King Herod — known widely in Rome and across the Empire as "The King of the Jews" - was a favorite of Augustus. Augustus appointed Herod Governor following Rome's annexation of Judea in 6 BCE. The Jews around the Sea of Galilee were less than enthusiastic about Herod. They took their Jewishness and their Hebraic traditions seriously, and were a source of continuing tension and occasional conflict with the powerful, ambitious, monument-building "King."

Herod maintained his residence and continued to govern the Roman provinces of Syria and Judea from the City of Caesarea Maritima on the Mediterranean Coast. The region around the Sea of Galilee was known to the Romans as Gallica, an administrative subdivision of the 10th Legion. On the death of Herod in 4 BCE (possibly the same birth year as the historical Jesus of Nazareth), Augustus chose to honor Herod's Last Will and Testament by dividing his authority among Herod's three sons, Herod Antipas, Herod Philip II, Herod Archelaus, and his sister, Salome 1. Herod Antipas was appointed Tetrarch of Gallica by Tiberias, successor to Augustus.

Gallica was bound on the north and west by the Roman Province of Syria; on the east by the Sea of Galilee and the River Jordan; and on the south by the Roman Province of Judea. Gallica included the tiny community of Nazareth, the probable birthplace of the historic Jesus.

Following the death of Augustus in 14 CE, at age 75, Tiberius, his adopted son, became Emperor.

In 26 CE, Pontius Pilate, a Roman bureaucrat, had succeeded the disgraced Valeris Gratus to become Governor of Judea, a province distant from Rome on the far eastern shore of the Mediterranean Sea. The Jewish capital of Judea – and the site of the Jewish Temple - is

Jerusalem. Pilate, however, continued to govern from the Roman capitol at Caesarea, with occasional visits to the Jewish capital to check in with the Jewish leaders among the three governing groups there: the Sanhedrin, the Pharisees, and the Sadducees.

Ziva calls out to her younger sister: “Gila! This way! Quickly!” Gila runs to her sister’s side to see what she is talking about. Around the corner, in one of the open shops to the side of the main path through the bustling market in lower Jerusalem, a young bearded man in a light grey robe is talking quietly to 10 or 15 people who have gathered round. Each in the small audience has turned an ear to the soft-words being spoken to them. This is the Jewish exorcist, healer, and preacher from the Galilee known as Jesus of Nazareth. The two girls had found him, by luck, talking with this small group.

The name Jesus is common. This Jesus is often known as the Nazarene to set him apart from the many prophets, exorcists, healers and aspiring Messiahs who work the crowds in Jerusalem during Passover. Because the preachers from the Galilee are especially vigorous and imaginative in their presentations, shoppers in the Jerusalem markets will almost always stop and listen if for no other reason than to be entertained on a warm spring day.

The name Jesus of Nazareth is more well-known than most of the others because his good reputation for healing and exorcism has preceded him to Jerusalem from the many in the Galilee who know of his work. Unfortunately, today, it is very hard to hear his words directly unless he comes close while moving among the listeners whom he seeks out to hear their concerns.

Jesus speaks of the one God in ways that fully engage his listeners. His voice is clear and modulated across a range of tones depending on the size of his audience, and on what he wants to emphasize to his hearer. He speaks to each person directly. He engages with them as individuals while they speak. He is alert to their questions, and thoughtful about his answer.
His eyes and those of this listener connect in a way that impresses an observer with the power of the bond between the two, even though it is a temporary one.

In spite of her excitement, Ziva is worried. She knows she should not have brought her younger sister to this place full of excitement and strangers without their parents to watch out for them. But today is special. It is spring with all the greenery and flowers bursting through the dusty earth; and it is Passover, a time to celebrate the emancipation of the Jews from slavery in Egypt 1300 years before. Their parents had come from Bethlehem to the main Jerusalem market in the lower city and to celebrate, with their friends, and with Jewish families from other Jerusalem neighborhoods, and nearby towns.

They had become separated from their parents, but Ziva knows how to find the house of the relatives her family is staying with. She knows her father, Avraham, a leather worker in Bethlehem, will eventually be found among the shopkeepers in the leather market. Avraham had alerted the girls that the Nazarene might be in Jerusalem during this Passover, even though the distance between Nazareth and Jerusalem is nearly 60 miles, a six to 10 day journey by foot.

Clearly his work as an exorcist, healer, and listener - but most particularly his prophesies – had begun to resonate beyond the shores and villages of the Galilee. Among the Jews of the Galilee, Jesus' eventual visit to Jerusalem had become inevitable, and this Passover had become the time for it.

Avraham described the Nazarene as a fisherman, carpenter, and day laborer. Some part of his work – as with all young Jewish men in the Galilee - involved work on the rebuilding of the Jewish Regional capital of Sepphoris. As construction work at Sepphoris neared completion, construction of the new Roman capital on the western shore of the Sea of Galilee began to speed up. Two years ago the architect and developer of the Roman capital, Herod Antipas, named the site after his friend, mentor, and patron the Roman Emperor, Tiberias.

Avraham knew these things because of the many weeks and months he had spent in the Galilee around that time earning good money as a leatherworker around the various construction sites. He had visited many small communities and had had many opportunities to listen to Jesus speak. Avraham worked with the fishermen from time to time as well and learned of the reputation of the Nazarene from these men of the sea.

Jesus had had no formal education and was probably illiterate. He started working at a very young age. He was very bright and curious, and he spent much of his time away from work at synagogue, learning and discussing the ancient teachings of the Jewish people with other Jews. Avraham had been among the other Jews on two occasions. He recognized early, the power of Jesus' presence and the clarity of his thinking even before he had learned how to articulate his thoughts in ways that brought an impulse toward faith to them, and to all within hearing.

Gila is also a very bright child with a driving curiosity that often gets her into trouble with adults as she pursues her many interests into things that might not otherwise be thought suitable for a child of 10. Of those activities most acceptable to her parents were Gila's extensive explorations of the streets and parks in her neighborhood. Once she was able to walk she would constantly stop and stoop to turn over rocks, to push away weeds so she could pick up bugs, and to dig around in loose soil to see what treasures lay beneath.

Gila eagerly questions nearby adults, even strangers, about the things she has discovered. When Gila presents them with a squirmy bug or slimy worm, some adults have shrunk away and insisted that she leave their presence, or have demanded that her parents, Avraham and Orli, retrieve her and take her away.

On this morning, their first in Jerusalem on this Passover, Avraham did not try to give Gila a sense of holiness and glory that he had felt in the presence of Jesus of Nazareth; a sense that Jesus could bring to any gathering; ...only to let her know that she might cross paths with him today. Ziva, a sister, a mentor, and a friend to Gila, takes all these interactions with adults, good or bad, in stride. The parents do not discourage either of their precocious children in their explorations and questioning unless there is a visible hazard to their safety or well-being. Their parental tolerance and their obvious regard for the Nazarene teacher has caused strains among neighbors and acquaintances of their household hosts who are otherwise inclined to be friendly, even to the two rambunctious girls.

This day is one among many in a period of generally tolerant, but occasionally tense relations between the Jews and the Romans. Conflicts break out over the question of the one God of the Jews, and the many Gods of the Romans. There were rumors that the new Roman Emperor, Tiberias – a far less competent ruler than his predecessor Augustus – does not deal well with contentiousness. He mostly wants to quiet the argumentative and vocal priests of the many Roman Gods among his advisors; ...particularly on the question of the one God of the Jews.

Tiberias, like Augustus, has no particular antipathy toward the Jews, but he has instructed his bureaucrats in the provinces to, once again, begin questioning Jews about their personal relationships to the many Roman Gods.

In response to these instructions, Roman soldiers were confronting Jews on the streets who were known to participate in Jewish holidays. The soldiers posed questions about the Jews and their relationship to the Roman Gods; ...Sometimes to one particular Roman God. The questions could provoke answers by individual Jews that might, or might not, be found to be acceptable to their interrogators. More often, depending on the social dynamics and intensity of the encounter, a soldier would ask about one of the minor Gods. Since the Roman Gods numbered in the hundreds, all the interrogatee could do would be to apologize for his or her lack of knowledge and promise to learn all that could be learned about the minor Roman deity.

By this simple pledge, most Jews were released from further interrogation. Some of the community leaders among the Jews are concerned, however, that some Jews were being taken into custody following these interrogations, and no family member or advocate is allowed to speak with them afterward. There are rumors in the market that a few are fearful that some are taken away to feed Rome's insatiable appetite for slaves to fuel the engines of its economy across its' expanding empire. Even more important among the upper classes in the City of Rome itself, are the constant needs for teachers for their children, and for household slaves with skills in cooking, serving, entertainments, and household administration. In these domestic skills Jews are often most favored.

When only a few listeners are surrounding Jesus, he often prefers to walk with them through the busy streets in Jerusalem during Passover. Many of the shop-keepers and their waiting customers wave and express their greetings to the popular preacher from the Galilee. Jesus walks with his hand raised to acknowledge all the greetings, and often tips his head toward a particular person when that person is known to him.

====

As Jesus begins to walk toward the market place, Ziva and Gila are torn between following the group, and leaving to find their parents. At that moment Jesus smiles in their direction, and motions them toward him. Jesus first asks after their language in the language of the Galilee, Aramaic. Ziva shakes her head to indicate that she does not speak Aramaic, but she asks, in the more common Greek, if Jesus speaks Greek? Jesus nods yes, but before he can say more, Gila announces that Jesus is a good Jewish man, so he must know how to speak Hebrew.

Jesus nods toward Gila then pauses briefly to gaze at her. He then speaks in a rough Hebrew that he struggles with. It sounds more like a second language when he asks their names and inquires after the parents of the two girls. He suggests that they try to find them while it is still light, as it is getting close to the time to eat the evening meal. Both girls nod. Jesus turns back toward his group of listeners and starts again to walk toward the market. In a last gesture, Jesus turns briefly to the girls and speaks of his hope that he will meet them again at this time and in this place at another Passover celebration soon.

They stand briefly, awestruck, saying nothing. Then they turn to find a leather seller in the market where they hope to meet up with their parents, and tell them the story of all their adventures, and their meeting with Jesus, on this wonderful day. They look at each other again, and then begin to run, slowly at first, and then as fast as they can, laughing and screaming through the narrow alleys and crowded spaces of Jerusalem's open air markets. As usual, their disruptions cause many bumped and ruffled adults to swear at them and condemn their parents.

Suddenly, as they turn to run down another alleyway, they see a group of Roman soldiers talking with one of the shop-keepers on the street a few dozen paces in front of them. Now, the soldiers and the shop-keeper are all looking at Ziva and Gila.

One of the soldiers appears to be in charge of four Legionnaires of lower rank. By the extra flash of red cloth on his shoulder, he is probably a Decanus of the second cohort of the tenth Legion. His name is Kanutus. He steps away from the group and walks toward the girls. The soldier speaks a rough dialect of Latin that suggests a poor upbringing, and lack of formal training in the language. Possibly he grew up on a farm near Rome. Since Latin evolved from Greek, Ziva is able to understand what the soldier wants to say, even though he does not say it very well in Latin.

Ziva and Gila turn and start to move away.

"Stop!" the Decanus yells, frowning "Are you the ones making all that noise?" Ziva and Gila, are both terrified, but both nod yes.

"Where are your parents?"

Ziva replies with the name of a nearby neighborhood known for the many Jewish families and store-keepers who live in the crowded apartments there.

"We will be in that neighborhood later today. What are your parents' names?" he asks.

Ziva's voice quavers as she quietly says the names: "Avraham …Orli." Her lip begins to quiver.

"Tell your parents that we will want to talk to them about your bad behavior. Now go along home, and don't run, or scream like the mean and disrespectful children that you are." The sergeant said.

Ziva and Gila walk quietly past the other soldiers, and past the shopkeeper who looks angrily in their direction.

Neither girl talks again until they are in their own neighborhood, and only a few steps from the shop-keeper's store and apartment where their parents are staying.

Orli turned from cutting carrots for dinner to greet them. "It is getting dark. I am glad you are home. Where did you go, and what did you do today?"

"We saw Jesus!" Gila shouted. "He was in the market with some of his friends. He talked to us!"

"What did he say?", Orli wanted to know. "Did he tell you why he is here?"

"He said he hopes to see us again during Passover!" Ziva chimed in. We must come back to Jerusalem next year!"

"Can we come back next year?" Ziva asked.

"We will have to talk to your father about that," Orli said. "I don't know how this visit is going for your father because the Romans seem to be very curious about whether we like their Gods, and which ones we like best. Your father tells them there are too many, and he gets confused. He always tries to convince them that they should think about the one God of the Jews." She paused and looked out the window. "I worry when he does that."

"Why, Momma?" Gila asks.

"Come help me peel these potatoes, my sweet ones. We will talk later about all of this with your father," Orli said.

"Bathsheva, their Jerusalem hostess and friend, walks into the small kitchen. She is carrying a bottle of wine; some olives, garlics, a tomato, and bread in her shopping bag. "Shalom," she says. "Is dinner ready?"

"I'd like to wait for Avraham. He should be here soon. We can have a glass of wine in the meantime. Where is Aharon?", Orli asked.

"He is at synagogue with some friends. They have been arguing over the meaning of the same part of the Torah for years; Bathsheva speaks in the lowered voice of a man: "'Does Avraham have the right to question God?' I am sure I do not know. After all these years of argument they don't seem to know either." She looks at Ziva by the stove, starting a fire to cook the soup. "Since Jesus is here, maybe he can clear up their points at contention; though I don't know what they will find to argue about after that."

"I have listened to Avraham discuss the two times in the Galilee when he and Jesus attended the same synagogue. I have not heard anything from Avraham that tells me that Jesus spends much time resolving points of contention.", Orli said. I understand that Jesus cannot read or write

Hebrew, but Avraham says he doesn't seem to have any difficulty picking out elements of flawed logic from the inflated ramblings of the Sanhedrin. But, …" Orli frowned, looked away, and shook her head. "I'm sure I talk too much."

====

The Decanus responsible for patrolling the public market in the lower city of Jerusalem this day, was told to stay outside the door while waiting to see his superior officer, Quintus, the Centurion Primus Ordinis of the Second Cohort of the 10th Legion. He had come, at Quintus request, to report on the movements of the charismatic Nazarene around the streets.

Decanus Kanutus was hand-picked by Quintus for duty in the Jewish capital of Jerusalem. Kanutus had been a decorated legionnaire who fought and was wounded during the purge of Jews from Rome in the previous year. He is loyal to Rome, to the Roman Gods, and to Quintus himself. Quintus knew Kanutus could lead a squad of Legionnaires into battle and would be effective as an enforcer of Roman Law in the contentious Jewish capital. When Kanutus arrived in Jerusalem, Quintus promoted him to Decanus.

Quintus also knew that Kanutus, once engaged in a face-to-face knife fight would not stop until he had killed his opponent. During the purge, Kanutus watched, helpless, as his brother was killed in a terrible way by one of the Jewish rebels. When Kanutus was able to break free of his own struggle, he followed the Jew, confronted him face-to-face, and, with a single, quick, slashing move with his pugio, spilled the man's guts to the ground, then left him to die in the hot dusty street.

Quintus is on the balcony of his temporary offices on the Mount of Olives. He is facing the setting sun as it cast the City and Temple before them above the Kidron Valley in growing shadow. He is speaking with Tribune Theodorus. Quintus motions Decanus Kanutus to join them. "Tribune Theodorus: please meet my decorated, and very able Decanus Kanutus. He has been down on the streets these past many days and has much to report. Perhaps, if you will give him your time and attention you can report back to Legate Kaius in Caesarea with much fascinating detail on our work here."

The Tribune turned toward the Decanus. Decanus Kanutus' knees were wobbly in the presence of these two legion officers. He knew Quintus from Rome and felt a powerful allegiance toward him. But he knew nothing of the Tribune, except, as is well-known among the ranks of legionnaires in the field, a Tribune is a political officer, not a fighting man. He could not be trusted under any circumstances. Since he would never appear on the field of battle, there was nothing that would change Kanutus opinion of this Tribune, nor of any other Tribune.

"Hail to Caesar Tiberias," Kanutus said, and snapped a formal Roman salute toward the Tribune.

"Very good, Decanus," The Tribune said, and returned the salute.

"Tell the Tribune what you have learned of the Nazarene," Quintus said. "Tell him the same things you have told me."

"I don't think the Nazarene is here to cause trouble for Rome. He talks to the people who know of him, or who have known of him because of his reputation as an exorcist and healer in the Galilee. He stays away from large crowds; he stops talking and walks away when too many people gather round to hear him. I think he will eventually be a bigger problem for the Jews in Temple than for us," Decanus Kanutus said.

"He is well-known, and well-regarded throughout the Galilee," the

Tribune said. "I have met him myself, and some of my staff have sought healing from him. The Legate is concerned about the effect he is having on population in the Gallica, and is very curious to know why he has come to Jerusalem for this Passover. Is this his first time here?"

"By everything I have found out about him, yes, it is his first time here," Kanutus said "but, I think he intends to return." "Why do you say that," Quintus asks.

"Because he told two little girls that he met in the market that he hoped to see them at another Passover soon."

"Are these little girls, or any Jewish civilians inclined to confide in you in some casual way when you are in uniform, Decanus Kanutus?" asked the Tribune.

"No, Tribune, but their parents - in hopes of avoiding trouble with Rome - are cooperative, and eager to please when they do not see an immediate threat. I knew that when I approached them in their homes at dinnertime later that day. I was discrete and spoke to their father as one father to another, concerned that certain bad behaviors among a father's children should not continue."

"What reason did you have for going to their homes?" the Tribune asked.

"Their two children, the two little girls, had been in the market and had met the Nazarene. In their excitement, afterward they began running and screaming through the streets disrupting the shops. I had a bit of business of my own in the market for some repairs to my uniform when they approached. I told them I wanted their parent's names, and they gave them to me. One of the shopkeepers told me where they were staying."

"It turns out," Kanutus continued," that their father, Avraham, an observant Jew from Bethlehem, has a reputation among the local Legionnaires for not responding to questions about the Roman Gods with enough seriousness. He is on a list of people who Governor Gratus believes need to be watched."

Quintus speaks. "Governor Gratus causes a lot of problems for himself with the Jews when he dismisses the chosen leader of the Jewish Temple and replaces them with his personal lapdogs like Ishmael. No wonder he is worried about Jews. I think Gratus may have partially solved the problem himself by accepting, so far, the appointment of Caiaphus." Quintus paused. "This sounds like a political problem among the Jews that will

take more time to either resolve itself or mature into something that Rome will need to be concerned about. But that time is not now, and you and I, Decanus, need to think about, and prepare for, our return to Rome.

"I will speak to my superiors in Rome, Tribune, about the problems Gratus has created for himself by his frequent removals and appointments of Temple Priests. As I said, it may have been partially corrected with his appointment of Caiaphas, but the actions of Gratus continue to need close watching."

Quintus continued. "I hope, Tribune, that you will inform the 10th Legion Legate of what you have learned here today. I hope, you will express to him, also, my belief that there is no immediate cause for concern. I'm certain that my Superiors in Rome will communicate any further information on this to Legate Kaius in Caesarea whenever it might become necessary."

"For now, please make yourself at home in my home, Tribune. Decanus Kanutus and I have some operational business to discuss, then we can all sit down to dinner.

After Tribune Theodorus leaves the room, Quintus speaks. "You have done very good work on this, Decanus Kanutus, but I think your work has only just begun. I want you to somehow arrange for me to get a closer personal look at Avraham and his family. Since you are now acquainted with him on a personal level, a chance meeting in the market might serve my purposes very well."

"This is a surprise request, sir, if I may inquire, what are those purposes?"

"You have told me that Avraham has spent many months plying his leather trade in the Galilee and is probably more well-acquainted with Jesus the man than most other Jews. I'm thinking it might be good to get a man like

Avraham and his immediate family back to Rome to help us educate ourselves and our Superiors about these Jews, and their beliefs.

Quintus paused and grew thoughtful. "You don't need to discuss this with anybody else, but I believe that their belief in one God will corrode the foundations of our faith, and our beliefs in our many Gods. Eventually the Jewish faith in one God will replace them. Rome needs to understand the political force these Jews represent, and must begin preparations for dealing with the consequences of their use of Jewish power."

Kanutus responds. "You may count on me, sir. I have to admit to a certain admiration for the Jewish belief, and the persistence of it among people who have little else of value that they can claim, though among my fellow Legionnaires, sir, those beliefs and the threats they pose to their personal Gods do not usually sit well."

"We will speak of this more tomorrow before you return to your duties," Quintus said.

====

The morning is warm on this sunny Passover weekday. Quintus is in a corner of the open-air square just off the crowded market. He sits at a table near a large shade tree sipping a small mug of wine. His uniform is casual. It is designed to show that he is a field commander in the Legion, but there is no obvious insignia of rank. Except for a short sword in his belt, he is unarmed.

He tears off a piece of bread, and dips it into a shallow dish of olives, oil, and spices. Quintus took note of a group of 10 or 12 ravens hopping around a piece of bread on the street a few steps away. One of the ravens took note of Quintus, and hopped sideways, in the raven way, toward him. Once able to secure the piece of bread from the table the raven hopped off. Eventually, Decanus Kanutus appears at the edge of the square. Quintus notes that the raven has a white feather on top of his right wing. With Decanus Kanutus is the subject of the previous evening's conversation; Avraham, the Jewish father of Gila and Ziva. Quintus motions them over to his table.

Decanus Kanutus approaches and offers a formal Roman salute. "Good morning, sir. May I present Avraham. Avraham was working with the leather smiths this morning. At my invitation, Avraham has agreed to come with me to meet with you."

"Very good, Decanus." Quintus motioned Avraham to a chair at his table. "Please sit down, Avraham. You may return to your duties, Decanus. I will find my own way back to the Mount of Olives. "

"Very good, sir."

"Please relax, Avraham. This is not an interrogation, and neither I, nor any of my troops mean any harm to you or your family. I do understand that some of the Legionnaires in the market are concerned about your casual attitudes toward the Roman Gods, but that is no part of the conversation I want to have with you now. I have always made it clear that I will not tolerate any complaints of harassment by any Legionnaires over conflicting Jewish opinions about the numbers or qualities of Roman Gods.

"So. For purposes of conversation, do we have at least this temporary understanding?"

"I feel very disadvantaged here. Why have you asked me to speak with you?," Avraham asked.

Quintus paused to gaze at Avraham. "I don't see any point in wasting your time Avraham. I will be leaving Jerusalem within the next two months, and I am looking for a Jewish family familiar with the Judean and Syrian Provinces, who will come to Rome with me, and work as household staff to myself and my wife."

"Why don't you simply take us as slaves during the dark of night? That is the usual Roman way, is it not?," Avraham Asked.

"Perhaps it will be necessary to take you to Rome as slaves, or so it will need to appear to others," Quintus said, "but that is not my preference. I can assure you and your family that you will be decently compensated while you work for me, whether as slaves or free. If you work well, we can discuss the return of you and your family to Jerusalem after some period of service, and I can personally guarantee some measure of status and some amount of compensation to help you resettle here again, or in Rome if that becomes your wish."

"You are making a very generous offer," Avraham said. "I have no idea how to respond to you, and I have no idea why you are making such a generous offer to a total stranger who, by his culture and upbringing may have hostile intent toward you, and everything you hold dear, and who could readily lay in wait for an opportunity to kill you, ...and your family."

"Decanus Kanutus said you were a reasonable man, but very sharp and very persistent. Let us leave aside the questions of motive, intent, and trust for now. Instead, let me describe the work I have in mind. I promise you, when the time is right, that I will tell you all that I have in mind as reasons for making this offer, and for making the offer to you in particular. First of all, I'm sure you realize that Romans in the upper reaches of Roman society often have extensive Jewish staffs in their households, and that they are usually treated quite well whether slave or free."

"Yes. I have heard of such things in Rome," Avraham replied.

Quintus continued. "My estate is just outside Rome on the way to the port of Ostia. I have been away from Rome and away from my estate for over a year. For now, my wife Cornelia is supervising others who are taking care of the properties, but when I return they will also return to their own properties. I need a staff to work the property right away. If I don't bring a staff back with me, I will need to recruit in Rome, and that will be very difficult, and very time consuming at that time of year. Worse, I fear that I won't find enough good staff to turn the soil and plant my crops. This possibility worries me a great deal.

"So. My reasons number 1 and 2; simple property maintenance and spring planting. Decanus Kanutus tells me you are a leather smith. I can assure you that you will have plenty of challenge for your skills in taking care of the 6 horses and tack on the property."

"So. My reason number 3; animal husbandry. Let me give you one more reason why I hope you will respond favorably to my invitation, then I will give you as much time as you need to ask me questions. Will that be alright with you?"

Avraham replied simply: "Yes."

"My reasons are simple. I see a worsening of tensions between Rome and the Jews in the coming years, and these can lead to destructive war. Do not misunderstand. I am a senior officer in a Roman Legion. If my Legion is assigned to go to war, then I will go, and will command my Legionnaires in ways that virtually guarantee that we will win. However, in peacetime, I will do all in my power to avoid war, and I will do all in my power to root out the seeds of war if I find that they will grow into war.

"So, my main reason is one of diplomacy. I would like to do everything I can to lessen these tensions so that we may both continue to prosper. As diplomacy, I need a much deeper understanding of the Jewish people, and I believe you and your family, over some period of time, can give me that."

Quintus looked hopefully toward Avraham. For his part, Avraham looked at the space between them, and said nothing for a few moments. Quintus waited.

"I don't disagree with your view of the tensions between Romans and

Jews," Avraham said. "But beyond the travel and the opportunity to see Rome as something other than a slave or prisoner, I don't see much of lasting value to me or to my family if I were to accept your offer, and I do see a great risk. Perhaps of most importance to me is my question to you about my four children; they are good children, and they are too bright to be happy peeling potatoes and emptying chamber pots at the direction of your wife no matter how pleasant she may be.

Quintus pushed the loaf of bread he had been dipping toward Avraham. "Remember that I will be paying you and the members of your family for the work you do for our estate. Your work with the horses will be well known among other horse owners in my area, and I will encourage those needing your skills to do business directly with you. Though I will want you to oversee the plantings, cultivations and harvesting, I also expect you to hire the members of your family and to recruit others when necessary to assist with the farm work.

"I am aware, in particular, of your two daughters, Gila and Ziva. They will both be in the care of my most trusted household staff, Kema, an Egyptian slave woman. Kema is with me looking after my household on the Mount of Olives. I will make sure you have an opportunity to meet her before much time passes. She is an educated and intelligent woman. She will keep your daughters challenged in all that they do, including the sciences and mathematics.

"Your son Lucius will work with the other Egyptian slave who is part of my household staff. His name is Anyim. He, too, is very intelligent, and is educating himself in his spare time when he is not repairing and replacing all the items necessary to the proper functioning of the house and properties. I am certain that Anyim will welcome some help from Lucius."

"I will make a guarantee to you now that your return to Jerusalem, with your family, is assured, though we will need to negotiate a date… perhaps 3 or 4 years into the future. When you return, you and your family will enjoy considerable status among your peers, in whichever ways you and your peers define that term. You have my personal assurance of that."

Avraham is thoughtful again. He pushes the bread back toward Quintus. After a few moments, Avraham speaks. "You have given me many generous guarantees. I know your title and status in the Legion, but I don't know your real status within the Roman army here in Judea, let alone your relationship to the Emperor Tiberias in Rome. If you lose their support while I and my family are under your personal protection, I fear the worst for us."

"I think I can arrange a secure way to get your family back to Jerusalem without my protection. Let us consider those possibilities as part of our larger negotiation. For now, I think we have laid out our concerns, and you need to begin a discussion with your wife. I would urge you to keep this entire discussion between us limited to the two of you for now."

"Is there some reason why you don't wish to break bread with me?" Quintus asks.

"Only that Passover forbids my eating of leavened bread. Nothing more." Avraham responds.

"Avraham continues, "Wait. There is one more important topic. You and your position in the Legion are well-known in the markets, here. If I return to Rome with you in apparent friendship and collegiality, I may never be able to return to Jerusalem with the degree of respect among my peers that you claim to be able to guarantee. In fact by the time of my return, any form of collaboration with Rome may condemn me – as it would with any Jew - to permanent exile, …or worse; …even death."

"Any guarantee among men is subject to the whim of the Gods, your one God, and my many Gods. The trip from Caesarea to Rome by ship will probably take about 21 days if we have good weather, fair winds, and strong oarsmen.

But there is no guarantee against hurricane force winds at any time on the northern Mediterranean Sea. Of course the chances for smooth sailing are pretty good for us in that time of year.

"The Gods are always willing to participate in the affairs of men, and, when they do participate, the first to go on the chopping block are often the guarantees made in good faith between men. The thing to keep in mind is that the Gods always need to be kept well-fed, happy, and unconfused."

Quintus continued. "I will leave you now. If you are contacted by Decanus Kanutus in the next several days, please pay careful attention to everything he has to tell you. Good day." With that, Quintus departs. Avraham wonders if he should have stood and shown some kind of respectful acknowledgement of Quintus' status relative to his own.

"Too late now," Avraham thinks to himself. He needs to get back to work in the Leather market, and he needs to figure out how to talk to Orli about all that he has heard today."

====

Avraham and his family go with Quintus to Rome. They had returned to Bethlehem after Passover. In early summer They had disappeared from the streets there in the way that many Jews had disappeared from the streets of Jerusalem following Herod Antipas' instructions regarding Jews and their relationships with the Roman Gods; ...quietly, in the night, and without a word to friends, or associates.

So far as anybody knew, even close family friends, Avraham and his family had been taken by the Romans; ...one of many so taken. As with these others, inquiries of the Romans by Jews about the whereabouts of Avraham and his family were met with silence.

On being told of their disappearance, a common response among Jews was: "Ach. I knew Avraham's making fun of the Roman Gods would someday ruin him, and his family. I am so sad for the children."

On the sea voyage between Caesarea Maritima and Rome, the weather had been fairly calm except for some contrary north winds near the Greek port of Rhodes. By his status in the Legion, Quintus was able to command space on a military Trireme with a larger sail, a keel and hull that had been lengthened to include more rowers and lightened by removing much of the battle gear to enable more speed. On one memorable attempt to set a record for fast travel between Caesarea and Ostia, a similarly equipped military Trireme made the trip in 11 days.

Though Quintus and his precious cargo was not trying for speed, his military attitudes as a Centurion - and, therefore, the superior officer on the vessel - were always in play. He kept the pressure on the oarsmen for a steady and deliberate rowing pace during windless days on their journey toward Rome.

With Quintus were his housekeeper in Rome, Kema, his Decanus, Kanutus, and two of Kanutus most trusted subordinates in his unit, Axius and Marcus. With Avraham were his wife, Orli; their two irrepressible daughters, Gila and Ziva; and their oldest child, Avi, and their youthful son, 12 year old Lucius.

Avraham's family, always mindful to keep up the charade of their slavery, nevertheless passed their time on the trip helping with cooking, cleaning, and loading supplies when in port. Lucius was particularly eager to help in running messages between Quintus and the Captain, and then the Captain and the rowing supervisors and sail pullers. When not running messages, he was ready with his full water bucket to go up and down the ranks of rowers several times a day, always eager to satisfy the thirsts of the sweating oarsmen on these hot summer days.

Most of the rowers were very grateful for the attentions of the children of Avraham.

The two girls were constantly asking questions of the rowers. They wanted to know everything about their lives when they were not on the vessel. Kema, kept careful watch over the girls when they went on these forays of inquiry among the rowers. Though the rowers on this Trireme were not slaves, and were being paid for their work, some of them had little patience for the curiosity and sheer noise generated by young children. Kema was quick to gather the girls up and carry them away at the first sign of a scowl on a tired rower.

Before finally agreeing to go to Rome, Avraham had worked out a basis for their interactions with Quintus which allowed Avraham as much latitude as Quintus was able to provide in assuring the security and comfort of all his immediate family. In order to provide the level of security that Avraham had requested, Quintus felt that Avraham and his family must always appear to everyone – even close friends and associates – to be slaves captured in Jerusalem. More importantly, they would always have to be ready to provide official paperwork to that effect.

The charge against them which caused their capture, essentially, was that Avraham did not show proper respect for the Roman Gods when questioned, just as Decanus Kanutus had originally reported. During this period of Roman authority over the Province of Judea a charge against Jews of not being sufficiently in awe of Roman Gods was a common one when senior officers in the Legion were to be rotated back to Rome. It was clear to all Legionnaires of any rank that Senior officers would, therefore, need to re-staff their households. Captive slaves were certainly a good start.

Senior Legion officers in charge of authorizing trips to Rome knew that someday they would also need cooperation in arranging travel for slaves when their own rotations home came up. For those reasons the documentation of official charges, and the creation of necessary travel documents for both senior officers and the slaves under their authority, tended to be routine.

Even with the full assurance by Quintus of the arrangements for their security and protection, Avraham could not help but be concerned about Quintus' frequent absences on Legion business in Rome. Nor could he put out of his mind the possibility that Quintus may get called to duty in some distance Province, and, terrible to imagine, he could be killed, or mortally wounded there. Avraham took some comfort in knowing that, though there were constant tensions between Romans and Jews, those tension were much less severe than they had been during the Jewish uprising following the death of King Herod some 20 or more years before.

If the terrible possibility of Quintus death on a distant battlefield were to occur, Avraham had no idea – despite the many assurances Quintus had given him about the loyalty and support of his family and household staff – what kind of problems he would have to deal with in getting his family back to Jerusalem. The problems seemed too numerous, and too impossible to imagine.

Quintus had promised to make arrangements for getting Avraham and his family back to Jerusalem if something were to happen that made it impossible for him to continue to provide security, but he had also said that he needed to familiarize Avraham with Rome, and the area around Rome before he could be sure about the success of a particular route and method, and a particular configuration of trustworthy friends and associates.

Beyond having as many conversations as possible with Avraham and his family on the Trireme, Quintus felt that travel around the city after they arrived home, followed by introductions to important and trustworthy friends of Quintus was the best way to work out a secure path back to Jerusalem should anything happen to Quintus.

Quintus had requested that Decanus Kanutus always be included in these discussions. Avraham was very pleased at this. His confidence and trust in both men continued to grow as he got to know them and know their thinking. The awareness of Decanus Kanutus of all arrangements and agreements between Avraham and Quintus gave Avraham an increased measure of security should anything happen to Quintus.

Quintus estate is near the small community of Vitinia on the road to Ostia a few miles west of Rome. Quintus had been as good as his word to Avraham that his family would be well-received by his household staff. Cornelia, wife of Quintus and overseer of the estate in his absence, went out of her way to welcome them and make sure they were comfortable in their new home.

Kema introduced Avraham's children – from the youngest, Avi; then Gila; then Ziva; and, finally, the oldest, Lucius, and finally their mother, Orli - to the household staff: Shani, an African slave girl who worked directly for Cornelia; and Anyim, a young African slave who worked with Quintus when he was on the property, or Cornelia when Quintus was not available.

After the introductions, Quintus motioned to Avraham, Anyim, and Lucius to follow him out to the horse stable. Quintus introduced Avraham and Lucius to the six horses, and to the arrangement of the tack and harnesses, the workshop, and the tools. Avraham could see that he would have many happy days working in this fine shop.

These good beginnings in the household of Quintus and Cornelia continued with only minor conflicts and upsets for several years. Gila and Ziva helped all the members of the household staff with their conversational Greek and Latin, and Orli worked with Cornelia on the general oversight of the sometimes rambunctious children and adults. When they agreed to make their time available, the male members of the household, including Quintus and Kanutus, also found themselves being tutored in language and reading by the two youngest girls.

Gila's intense curiosity was greatly rewarded when she discovered the Household library in a corner room of the house that opened to the south. Gila had learned to read early whenever she could find someone to teach her, but she knew her ability to read was rudimentary. This library, and Cornelia's promise to help her read any of the scrolls in it, left her awestruck. She was barely able to contain her impulse to pull a scroll off the shelf and begin reading at once.

Ziva is also a reader, though somewhat less intense about it than Gila. Ziva's enthusiasms went more to writing and the use of calligraphy in the forming of the letters of the Greek, Roman, and Hebrew alphabets. The library contained so many scrolls and maps, that she immediately realized that she could not master all of the writing styles no matter how long she remained in the Quintus household.

Cornelia worked with frescos and mosaics. She involved Orli in all aspects of the several arts projects in these two mediums that Cornelia had been developing for the house. Orli showed herself to have a lot of natural talent in casting the imagery and selecting and mixing the colors to be used In the projects. She began to see a future for herself in this kind of creative work.

The four children of Avraham also grew in their awareness of the world, and in their ability to focus on those things around them that drew their interests. They were aided in this by their warm and loving parents, and by the intellectual stimulation and welcoming atmosphere created by their two Roman hosts, Quintus and Cornelia.

The three African slaves – Kema, Shani, and Anyim – also prospered by their close proximity to Avraham's family members, particularly the children, and most particularly by the training in language conducted every day by Gila and Ziva. Training in everyday Greek was constant with reminders, penalties, or rewards, as needed, and at any time of the day, and as determined by the two girls. Gila was very pushy and often intolerant of errors. Anyim, growing quickly into a mature African man, often became frustrated by this abusive treatment by a child, and would occasionally stomp out of the room. After a period of cooling off, one or the other of the two combatants would extend a hand of friendship, and the learning would then resume.

As time passed training in the spoken languages of the surrounding community were expanded to include lessons in writing for all of them by Ziva, and by special tutors brought in for the purpose by Cornelia.

Avraham maintained his part of the bargain with Quintus. Several times each week through the year, and more frequently during important Jewish festivals, Avraham would sit down with Quintus to discuss one of the many topics that Quintus wanted to know about. In return, Quintus took Avraham with him when he went into Rome on either Legion or personal business. Quintus introduced Avraham to his professional friends and business associates as a slave, captured with his family in Jerusalem. After a pause, Quintus would always add: "… and a very valued member of my household staff."

Avraham's work with the horses lead to several opportunities to compete with other horse owners in his neighborhoods and in Rome for selection of two horse teams to pull chariots. Chariot races were held regularly in the Circus Maximus a little way south of the Roman Forum. These intense and exciting races provided the final test of a horses' ability to carry a driver and a Centurion, or a driver and a pair of archers into battle.

In most of the horse community, this was much of the purpose of raising horses on the rural estates outside the City of Rome. The Legion paid well for good horses, and Quintus was able to earn significant amounts of money from the horse husbandry that Avraham was able to bring to his estate. The development of a horse stable capable of selling good chariot horses to the Legion could not have occurred without the attentive care and patience of Avraham.

=====

In the winter of 26 CE, Quintus announced that he is being called back to Jerusalem to work with the Jewish leaders to try to resolve some conflicts that were becoming difficult for the Roman political administrators in Caesarea. This would be a diplomatic mission. Talk among the leading Romans and Jewish leaders was needed, not military threats. The Cohorts of the 10th Legion, which Quintus was also fully able to command would – if called into action - introduce additional and, it was felt, unnecessary tension to some already tense discussions over matters of importance to the Jews.

As always, particularly since the days of King Herod, the increasingly tense points of conflict had to do with the many Roman Gods, and the one God of the Jews. Though the chances of resolving this fundamental issue over the long term were poor, the relatively new Roman Prefect, Pontius Pilate, a practical man, hoped to keep the conflicts tamped down for as long as he remained Governor. Quintus seemed especially well-equipped to handle this kind of assignment because he had an excellent reputation as a very competent officer in battle, and he did not have a reputation among the Jews as a brute and a murderer, unlike many others among the Legion's officers.

With a sign-off by the Prefect, the message went out to Quintus that he should gather his most trusted staff and come immediately to Jerusalem. He should schedule himself to arrive a few weeks before the first day of Passover. Unfortunately, that would mean sea travel in late Winter when weather on the Mediterranean could be very bad.

====

Shortly after the Roman mid-winter celebration of the solstice, Quintus called his household together to begin discussions of the return to Jerusalem.

"I have asked you all to join me to discuss a matter of considerable importance to all of us, though it is of particular importance to Avraham and his family," Quintus began. "The Governor of the province of Judea Palestine, Pontius Pilate, has asked me to gather my staff, and others who can support my work, and come to Jerusalem to help resolve some emerging conflicts between the Jews and the Romans. These conflicts need to be resolved soon, because they have the capability of escalating into situations that can be violent; … situations that none of us want to contemplate. Pilate and I want to do all that we can to bring about as many resolutions as we can."

"I will, of course, take Decanus Kanutus and his two Legionnaires Axius and Marcus as they have always been the essential part of my most trusted military staff. I do not know how long I will need to stay in Jerusalem, so I will take Kema because I have total confidence in her ability to maintain my household in that distant city.

"Shani will stay here to continue working with my good wife, Cornelia. Anyim is now a grown man and fully able to manage any part of the work of this estate. I have not had an opportunity to talk to Anyim about any aspect of this yet, but his future – whether here running the estate with Cornelia, or with me as a trusted member of my Jerusalem staff is very much up for discussion." Quintus paused a moment, then turned to Avraham.

Quintus began again: "From the day I first met Avraham in Jerusalem, we have discussed the day when he and his family must return. That day is now rapidly coming, and it is time that we all hear his thoughts on the very important decisions that are coming upon us."

"You have not said how many people will be allowed to accompany you," Avraham said.

Quintus responded. "Pilate sent his formal request to me by personal courier. The official document was signed with his own very distinctive signature. Along with Pilate's signature, there were several signatures of my colleagues in the 10th Legion. I don't think anybody will refuse to honor whatever request I might make in terms of the people I chose to bring with me."

"So, you are saying that I can bring my entire family back to Jerusalem?"

"Yes. Though, I think you need to discuss this with each of your family members before making your decision. Your wife, Orli, has become a very accomplished fresco painter and sculptor; Lucius is now a grown man, and has gained many skills that are very useful, and valuable, in horse husbandry and maintenance of an estate of this size; Ziva's teaching skills in reading and writing are very impressive. These skills are valuable anywhere within the Roman Empire, and their language abilities, and their ability to read and write mean that they can make their way anywhere. I would ask each of your children to consider whether they want to pursue their futures here, in the heart of the Roman Empire, or in the distant provinces of Syria and Judea.

Orli spoke up. "My children, except for the youngest, Gila and Avi, can make up their own minds on their futures. I will go where my husband goes."

Quintus thought to himself for a few minutes. He rubbed his chin as he often did when confronted with a difficult and tangled problem that needed to have some kind of a resolution set in motion. "I want to speak about Gila. She is now an almost grown woman, and a very attractive one at that. But her real beauty is not only in her pretty face. It is in her mind. In addition to her constant work trying to bring all of us up to some unreachable standard of literacy, I have taken note of the time she spends in the library looking at the scrolls and making notes to herself.

"A few years ago I first saw Gila at work in the library. I went in to sit with her for awhile and asked what she was learning." Gila is sitting quietly off to the side of the group. As always, however, her eyes closely follow each speaker as though she is fascinated with whatever is being talked about. Quintus looked at her. "Do you remember that day, Gila?"

Gila blushes and looks at Quintus. "Thank you for those very kind words, Quintus Caecillius. I do remember the day. I was working with the papyrus roll copied from one of the works of Cicero about the workings of the Roman Senate. I was comparing his thoughts with some of the writings in Torah. My dad said this might be a useful thing to do." I was fascinated with it. I went on to make several more of these comparisons. I now have a lot of notes. Should I read one of them in Cicero's original Greek?"

Gila looks around the room. She burst into a big smile that wrinkles her nose, opens her mouth to show her teeth. Deep dimples form around the corners of her mouth.

Avraham laughed. "Gila has always had a good sense of humor." Others

around the room also smile at Gila's joke.

Quintus motioned for quiet. "I hope everyone understands that I very much want Gila to come with us to Jerusalem. Her erudition and charm will work miracles on the contending parties there. I can't imagine trying to do my job without someone as talented and as competent. But it is up to Avraham and his family to make the decisions on which of his family, besides himself, will come with us."

Several similar meetings took place over the next few months as each member of the household made their personal and collaborative decisions on staying on the Estate, learning more about the workings of the Roman Empire in the pinnacle of its' power, or facing all the possible risks of a trip into the troubled piece of ground in a far distant province known as Judea, and to it's capital; the capital of the Jewish world, the Jewish temple, and the city of Jerusalem.

Finally, it was agreed that Avraham would take Gila and Lucius with him. Orli would stay on the estate. Neither she nor young Avi would have much to contribute to the likely mission to be assigned to Quintus. At the same time she had much to learn about the visual arts and the wonderful and colorful clays and stones that she had become so adept at working in the creation of beautiful objects.

====

The sea voyage was mostly without incident except for some stormy weather, and stiff, choppy, northerly winds across most of the Aegean Sea in the crossing from Corinth to Rhodes. Pilate was eager enough to have Quintus in Jerusalem that he had requested one of the faster Triremes to transport him. These ships combined space and comfortable accommodations for high level Roman officials; extensive storage to cut down the number of provision stops on the journey; and, finally, specially selected rowers and sailors who were young, strong, and eager to impress the Roman officials who needed speed no matter what the weather. Crew experience with changing weather conditions on the well-understood 20 day run between Ostia near Rome, and Caesarea near Jerusalem also counted for a lot.

Even with a strong crew, the choppy winds in the Aegean could cause a lot of rocking. Rowing became difficult. Using the sail to power the ship became difficult and dangerous on the rocking and surging deck. The guests of honor on the boat – Quintus, Decanus Kanutus and his two Legionnaires, Avraham, Gila, and Lucius – were advised to stay in their quarters below decks at the rear of the ship for their own safety. Initially, Gila and Lucius were both disappointed that they could not mingle and work with the rowing and sailing crew as they had done when Quintus first brought them out to Rome from Judea.

After a few days in enclosed spaces on rough seas, the entire party, even Quintus and Avraham were glad they had no immediate responsibilities. Quintus was the Ostensible ranking officer on the Trireme, but the ship's Captain in charge knew that he had a free hand from Quintus to deal with the weather and the crew however he saw fit.

Within the overall need to get to Judea as quickly as this fine ship could carry them, the rough and variable weather meant that decisions were often required on directions of travel and ports of call, when changes in schedule were required. When necessary the Captain came to Quintus, outlined the problem and his proposed solution, then answered whatever questions Quintus might have.

The biggest problem during a long spell of rough weather was the growing fatigue among the rowers and sailors. None of the deck crew were slaves. All were Legionnaires who are assigned full time to the ships. All the crew had fought to get assignment to seats when they became available on the ship and were very competitive in keeping their seats when these special Triremes depart on their next voyage.

Though discouraged by the Legion, bribery was occasionally used to help gain seat assignments. Ultimately, though, an oarsman's ability to keep his seat in successive trips depends entirely on the Captain's assessment. The stakes are high. Successful completion of their assigned trips means generous bonuses, seat assignments on future trips, opportunities for promotions, and better onshore accommodations in whatever port they find themselves.

In weather with strong, but stable winds, and with the sail up, the rowers can rest. In calm weather, the rowers work in shifts, usually with half resting while the other half row. In stormy weather with squirrely winds, or when going into battle, all rowers take to the oars, and all sailors make constant adjustments to the sails. The constant shifting of personnel because of fatigue becomes a regular subject of discussion and decision making between the Captain and Quintus during the Aegean crossing.

As always Quintus is inclined to allow the Captain full authority over the trip, except when the Captain has to recommend that the ship be taken to an area of calm where repairs can be made, and the crew given a period to recuperate. Quintus knows that Pilate is eager for his arrival, and any unnecessary delay will not go well with either the Captain, or himself. He allows the Captain one unscheduled stop for crew rest, and ship repair at the Isle of Rhodes, a port on the eastern end of the Aegean, just south of Rome's Asian Province.

Fortunately, the winds to the east of Rhodes abate, and the remainder of the trip is relatively calm.

=====

At Caesarea, Quintus and his party of Avraham, Gila, Lucius, and Decanus Kanutus and his two Legionnaire Guards are met by a welcome party sent by Pilate. To Quintus, the welcome seems warm and genuine enough, although the delegation spokesman, a Tribune named Ephesius, suggests that they take a few days to get rested and settled, then plan to meet with Prefect Pilate mid-morning Friday. "The scale of events is growing more intense. Passover is coming soon. The contending parties need attention, and we need some of your skills in diplomacy. Are you and your staff up to it?" Ephesius asks.

"I have every confidence in my staff. If you will give us a chance to get our feet on the ground, and get cleaned up and organized, we will be ready to discuss our plan with the Prefect," Quintus replied.

Three days later, Quintus and his private staff of Avraham, Gila, and Lucius, arrive at the Governor's offices to meet with Prefect Pilate. Decanus Kanutus and his Legionnaires have been detailed to acquaint themselves with the Senior Legionnaires at his level of rank in order to give Quintus a report on the likelihood of receiving their full cooperation if and when needed.

Quintus notes that, as usual, Pilate likes to keep important visitors waiting as a reminder of his authority over the agenda and schedule of whatever meeting is to occur. Senior Legionnaires like Quintus recognize the need to tolerate these delays when there is urgency in the topic; Pilate has a nasty habit of abruptly cancelling important meetings if he becomes offended by an action of one of the participants. Sometimes he does not re-schedule the meetings. In this case, however, Quintus was sure that Pilate wanted this meeting more than he did.

Eventually, Pilate welcomes them in. After an exchange of greetings, Quintus, Avraham, Gila, and Lucius sit down on one side of the table, and Pilate and the Tribune who had greeted them a few days before, sit down opposite.

Quintus waited for Pilate to open the discussion as is dictated by protocol.

"Two of your staff members seem very young Quintus," Pilate said. Do they have the skills and experience necessary for such an important mission?"

Quintus replies: "The two youthful members of my staff, along with their father, Avraham, have been slaves in my household since their capture in Jerusalem nearly 8 years ago. They have lived and worked with my family in complete harmony for that whole time. They have immersed themselves in Roman culture and attitudes. They have learned and have become adept at all the languages of the Roman Forum and marketplaces.

"As you know, I have an extensive library of writings by prominent Roman, Greek, and Egyptian writers. Gila, in particular, is now fluent in those languages, and she has become learned in many of our most important writers. She has been developing a comparison of the Roman Gods, with references to the Jewish Torah based on her researches in my library. She is young, but I believe she can hold her own in any discussion of nuance between Roman and Jewish philosophers. "

Pilate looked at Gila, then, after a pause, again at Quintus. Pilate's reputation for occasional lechery was on Quintus' mind as he waited to hear Pilate's next question. "So, Quintus. A very solid endorsement of this very beautiful young woman. Avraham and Lucius? How do you see their role in any mission I may assign to you?"

"Avraham has been taking care of my horses. The horses are bred for sale to the Legion. Avraham has developed a very fine breeding line of such horses. He works with other horse wranglers in my area west of Rome to develop racing horses for contests in the Circus Maximus. The horses bred under Avraham's husbandry consistently win those contests. I believe that whenever he is called upon to share his knowledge either with your horse handlers, or with any other Roman horsemen or handlers we might come in contact with will have very favorable results."

"And Lucius. What special talent does he bring?"

"Lucius has been responsible for the day-to-day workings of my estate, and for making sure that everything is in good working order so it is always ready when needed. He works closely with my wife Cornelia, and with my household staff in taking care of purchasing and storing items needed by the household. He works with his father, Avraham in taking care of the horses, and with his sister, Gila, in helping with the language education of all the members of the household. I consider him to be indispensable," Quintus concluded.

"There is one last note that I want to pass along to you, "Quintus said to Pontius Pilate: "Avraham, Gila, and Lucius are Jews. They are born Jews and have been Jews all their lives. I have not interrupted their study of their books, and I have not prevented them from attending Synagogue in Ostia. This is not favoritism on my part. My two other household slaves are Nubian from Egypt. I do not interfere with the practice of their religion either.

"For myself, my wife, my children, my household staff, and the Legionnaires on my personal staff as Centurion Primo, we are Romans, and we live by the Roman Gods. I encourage all the members of my household to observe as we pay obeisance to our Gods, which they sometimes do, and they invite us to return the favor, which we also sometimes do."

Pilate is abrupt. With a barely concealed sneer, he interrupts Quintus "That is all very noble and generous of you Quintus. You have convinced me of the qualifications of the people you brought with you for this assignment. To have brought such accomplished and knowledgeable Jews will no doubt be of interest to our chief Jew, Herod Antipas, the Tetrarch, and may help gain his support for whatever you propose to do. …but my patience grows thin, the day grows long, and my reasons for bringing you to Jerusalem have not yet been introduced.

"The Jewish Passover will be upon us in about six weeks. It is important that you and your staff be there for the weeklong celebration. Over the past few years, there has been a growing stir east of here, in the Cities of Sepphoris, Tiberias, and in the farming areas and villages around the Sea of Galilee.

"This particular stir – and the reason I brought you here - concerns a new, badly dressed prophet of baptism, and a new movement that has grown up around his preaching. While I am usually not much concerned about these minor sects that grow, live, then soon die outside Jerusalem, I am concerned when the Jewish establishment in Jerusalem gets all excited and begins to describe one of them as the new Messiah.

"The Tetrarch, Herod Antipas, has had the most prominent of the preachers, John the Baptist, arrested, but not before he had baptized one of the other well-known preachers out there, Jesus of Nazareth."

At the mention of the name Jesus of Nazareth, Gila eyebrows furled, but she quickly restored her face to the submissive state she is trying to maintain in this meeting. Fortunately, Pilate was turned toward Quintus, and did not see the change in Gila's facial expression. Quintus, however, took note of the subtle shift in her demeanor.

Pilate continues: The legionnaires in these rural areas have told Herod that an arrest of another one of these baptized preachers, particularly the Nazarene, will generate a lot of anger among these Jews around the Galilee. This is a growing concern.

"The Jewish 'nobility,'" Pilate says this with an arched eyebrow and rolled eyes "...the Sadducees. Pharisees, and the Sanhedrin – are also concerned about this 'baptism,' because the idea of total immersion as the only way to wash away sin and corruption does not conform to their most cherished beliefs. They are equally concerned about this Jesus, the Nazarene, the most prominent and commonly-known of John's baptisms

"So far, I have said that I have no plans to interfere with Jewish teaching or teachers, but I cannot speak for Herod, or for Phillip, the Tetrarch of Gallica and Syria. If the Jewish leaders are concerned about the welfare of John the Baptist while he is being held, I have told them that I will carry a message to Herod about their concerns. I will recommend that he be released to your custody. If that is not possible I will strongly recommend that they make sure that he does not come to harm while in their custody.

"To me it is either make all the religious sects in the Provinces, of Syria, Gallica, and much of Judea angry as hornets, or make a rather small and clubby group of Jews who believe themselves to be infallible understand that they have no monopoly on the practice of faith, especially when it is their own faith they are worried about.

"So, Centurion. As you can see, the situation is complex. I want you and your staff to come up with a course of action for determining what to do about this. For the next two weeks, you will have whatever military resources you think you need – within reason - from among your fellows in the available cohorts of the 10th Legion; your experts – particularly Gila – will have as much access as she feels she needs to our Libraries, and to our best local thinkers and observers of the situation here.

For the next two weeks after that, plan on being in Jerusalem with all your civilian and personal staffs, and plan on staying at least another week beyond. Depending on your findings during Passover, I think we will want to meet here again to decide the next step in our plan. Questions?"

"No Prefect. We have all we need. We will take our leave and check back with you on travel arrangements to Jerusalem In two weeks, if necessary. Of course, I am available to you at your request any time."

"One thing more, "Pilate said. "I'm told Avraham is from the Sea of Galilee. Is that correct Avraham?"

Avraham replies: "No Prefect. I was born in Bethlehem, as were all my family. I and my family had a leather working business in Bethlehem before being taken as slaves. Before that I spent several years working on the re-building of the temples and monuments at Sepphoris, and on the construction of several structures in the new capital of Tiberias, on shore of the Sea of Galilee."

"Would you be comfortable returning to the Galilee, and re-acquainting yourself with past friends and co-workers?" Pilate asked.

"Yes," Avraham replied.

Pilate looked at Quintus, "Please consider sending Avraham into the Galilee as part of your plan. Questions?"

All of Quintus' party rose, gave a proper salute' and departed.

=====

The Jewish Passover in the modern calendar year 30 CE, begins with the Passover Seder at dusk on the first day. The Seder is a festive event full of eating, drinking, and ritual. As part of their education process, Avraham, Lucius, and Gila had come down to Jerusalem from Caesarea to reconnect with their friends from years ago, Bathsheva and Aharon. Avraham hoped that Bathsheva and Aharon would agree to invite Centurion Quintus along to the Seder, and to other celebrations and rituals that would take place during the week of Passover.

Avraham explained their new situation of slavery to the Roman Centurion Quintus, and they explained what they were now doing on behalf of the Centurion Quintus and the Roman Prefect Pontius Pilate. He explained that Pilate himself had brought Quintus back from Rome, and that Quintus has brought Avraham and his family along as indispensable staff.

Avraham had complete trust in Bathsheva and Aharon, but he knows that the issues and problems he is telling them now will test that trust in ways which none of them would ever have anticipated in their many years of friendship.

"You must trust me, Aharon, when I tell you both that Quintus bears no ill will against any of us here, or against any of the Jewish people. Based on our family's several years of intimate service in his household – I believe he, and the Prefect, are working toward a long-lasting peace that will include all of Syria, Palestine, Gallica, and Judea. This is in the long-term interest of all Jews, and all Romans. "

"I would trust you with my life, Avraham, "Aharon said.

Bathsheva also spoke: "I share my husband's feelings, Avraham. I only

wish your entire family were here with us. I am overjoyed to see what a handsome and intelligent young man and woman that Gila and Lucius have become. If possible, can you have Quintus visit with us before the start of Passover so we will have a better idea how to introduce him to our friends at the various events. As I am sure you can imagine, there will be many questions. We must be open to whatever is said, and we must be willing to fully answer whatever questions are asked."

"I'm sure you realize, Avraham, that there is an increased level of tension in Jerusalem because of the capture of John the Baptist by Herod Antipas several months ago. The longer he remains in prison, the more this tension increases the risk of violence during Passover. I am also told that the Nazarene, the most prominent of those Baptized by the Baptist, will be coming to Jerusalem from the Galilee, and may already be here."

Gila could not contain her excitement at this. "Oh, Father! I must go to meet him! I know him, and I am sure he will remember me from the meeting Ziva and I had with him on the streets here several years ago."

Avraham smiled as he put his hand on Gila's shoulder. "I don't know if I have told you that Quintus and I have been discussing such a meeting, Gila. It is potentially fraught with difficulty and risk. Jesus is probably a much different person than the one you met. The cult of Baptism is a challenge to the establish rules of worship by the Sadducees and Pharisees, and they are very sensitive to the possibility of further degradation of their authority.

"Depending on the setting and circumstances the Nazarene is the kind of speaker who can turn a simple discussion of faith into an incitement to violence. Both Quintus and I are very concerned about this. It is contrary to what Quintus and Pilate are trying to do here, and, more than any other consideration, I fear for your safety."

"Do you Trust me father?" Gila asked.

"Yes. Of course, Gila."

"Then let me work out a plan for how such a meeting might take place with safety, and how it can be made to serve the needs of all of us. Lucius? Will you help me? Will you agree to go with me to whatever meeting I am able to work out with the Nazarene?"

"Of course, my sister. I will follow you wherever you go, and I will protect you with my life."

Avraham and their hosts seemed to be struck silent by this expression of confidence and faith by these two children. Avraham knew that he must say yes to Gila's request.

"Can you be ready to discuss your plan with Quintus and his Legionnaires in three days?" Avraham asked.

"Yes, Father."

Avraham rose to gather his family and thank his hosts for their hospitality. He asked, softly, if they might be willing to work with Gila depending on what she may propose.

"Yes. Avraham. Of course. You may count on both of us to do all we can in the interest of lasting peace."

Bathsheva nodded her assent. She looked deeply into the eyes of each of them in turn. She smiled slightly as each looked directly at her in return.

=====

Gila proposed a meeting with the Nazarene close to one of the main streets in the lower Jerusalem marketplace, but at least partially isolated from the crowd that would flood all the streets of all of Jerusalem on this precious and popular holiday.

Quintus consented to these arrangements. He would attend the meeting with Gila. They would invite the Nazarene to bring along whomever he might choose. Discussions will take place in the language the Nazarene is most comfortable with. If the chosen language is Aramaic, Quintus and Gila will bring an interpreter skilled in the nuance of that language. If the Nazarene prefers a language he is not totally comfortable with he may bring also his choice of interpreter to assist him.

The purpose of the meeting would simply be to explore issues and problems of concern to the Nazarene, with the promise that Quintus and Gila would carry his concerns to the appropriate senior Roman and Jewish officials.

Other meetings with the Nazarene will follow. Quintus would have a small contingent of no more than 10 battle-experienced Legionnaires with him. They will be in uniform with battle gear. They will be moving casually, but together, and out of marching order in full view of passersby in the market. They will be ready to act quickly, and with deadly force if necessary. They will be trusted men who fully understand and support this initiative. They will give their lives to it if that is what is required.

=====

Jesus the Nazarene agreed to all these proposed arrangements when he attended a dinner at the home of Bathsheva and Aharon a few days before the Passover Seder. Bathsheva and Aharon were both held in high regard with most of the Jewish factions in Jerusalem. In large part their good standing – and their success if persuading the Nazarene to come to their house - was a product of their good works in the Jewish community, and their long association with Avraham, who had spent so many years working in the hot-bed of religious tumult around the Sea of Galilee, and the charm of Orli, his wife, and all their wonderful children.

When introduced to Gila, the Nazarene was very pleased to make her acquaintance again after many years. "I knew we would meet again, Gila, though I did not think it would be so long. I hope our time together will be profitable for all parties. With your full participation, I have no doubt that it will be."

As the meeting broke up, the Nazarene asked Avraham if he could have a few private minutes with Gila. Avraham agreed.

In the shadows outside the house of Aharon, Jesus spoke quietly in Aramaic: "I cannot tell you how pleased I am that you are involved in this Gila. I knew when I first saw you that you have a great intelligence that will be essential as we Jews continue to work toward religious and cultural independence and security within the Roman Empire.

"But you must be careful. I'm sure you know by now that the leadership of the Romans and the Jews have a great capacity for deceit; who may say many things for effect, but who have very simple goals: the Pharisees and the Sadducees do not want to see the Baptists gain influence, because they are a clear threat to the religious authority of the Sanhedrin. The Romans probably don't care how the Jews sort themselves out, but they have no interest in losing control of the situation in Judea, Syria, and Palestine. The Romans will sacrifice whoever is necessary on whichever side to insure that result.

"I do not want you to be hurt, because I believe you, and Lucius, and your family, and even Quintus and his family are very much at risk in this process that Pilate has set up, and I believe you are all very important to the future of this part of the world."

"Your words take my breath away," Gila said. "I feel a great burden, and I am frightened that I will not be able to do what I am being called upon to do. Can you help me?"

"I will be with you whenever you and your friends and family are among my people. For various reasons, I cannot be with you when you are among the ones who fear me, nor can I be with you among those who have called me 'enemy.' In those times you must rely on the word of Quintus that he and his chosen Legionnaires will protect you from physical harm and will protect all of those you hold dear. I have looked into his eyes on many occasions, and I believe him to be a good man."

"How do you know Quintus," Gila asked. "Have you ever met him?"

"The same way I know you." Jesus paused. "I must ask you to do one last thing for me."

"Of course. Anything you ask."

"I have come to believe in the power of fully immersing the body in the baptismal pool. It is only in this washing away of all of the sins and corruption of the body that redemption and rebirth is possible. This is what I have learned from my own baptism by the hand of John the Baptist. I believe this, and I believe that all who have followed the example of myself and John also believe it. "

Gila began to feel a little uncomfortable with the Nazarene. What was this leading up to?

"Do not be disturbed, Gila. I know you and all your family are devout Jews, and you have now become a very learned Jew. I only want you to realize that there are new ways of expressing devotion to God, and I want you to be aware of the one way that I am most pleased with. If you ever decide to be baptized, please be assured that I will want to conduct the ceremony. Will you keep me in your thoughts?"

With that Gila leaped into the arms of Jesus and hugged him very tightly. She had not had these kinds of feelings before, and she was a little shocked and amazed at them, and what they had just made her do. For his part, Jesus hugged her back, for he was not above the sins of the flesh, though he had no desire to take advantage of this beautiful child.

He did, however, turn her face toward his and kiss her full on the lips. Gila pulled slowly and carefully away from him. She continued to look into his eyes. Jesus smiled, then turned away and was gone into the night.

====

Pilate's plan had worked, at least for the time being. Passover went without incident, and Jesus of Nazareth was able to move freely among all those who observed Passover with varying degrees of devotion and emotional intensity. As promised, Quintus had taken the concerns of Jesus to Pilate, and Pilate had given every assurance to Quintus that he would take them up with the Tetrarch, and with the other powers-that-be in the empire just as soon as it was appropriate.

The party that had traveled to Judea with Quintus all returned to Rome. Avraham discovered that his work with the horses in Quintus stable was far more interesting to him that anything that might happen during the years left in his life in Jerusalem. Lucius also felt his future lay more in Rome than in Jerusalem.

Gila returned to her now well-trodden path of scholarship and diplomacy. Quintus made sure that she was enrolled in the finest teaching institutions available in Rome, and that she was given full access to all of Rome's philosophers and statesmen. He also made sure that she would have opportunities to travel to the far corners of the empire to learn local languages and custom, all part of her training as a high level diplomat with a specialty in the religions and languages of Rome's eastern provinces.

For her part, Gila was pleased with all of this, but her mind did often wander to the night in Jerusalem when she had been kissed by a man considered by all to be the Jewish Messiah.

PANEL 5 - COLLAPSE

Agna, a young woman, works as a helper in a small bakery near the docks of Hamburg, Germany. Suddenly a cry of plague! rings through the streets. She grabs her small bag of belongings, and her few precious bits of parchment, pen, and small bottle of ink for writing. She tries to escape the Black Death spreading from the docks by a cargo ship carrying Bubonic Plague.

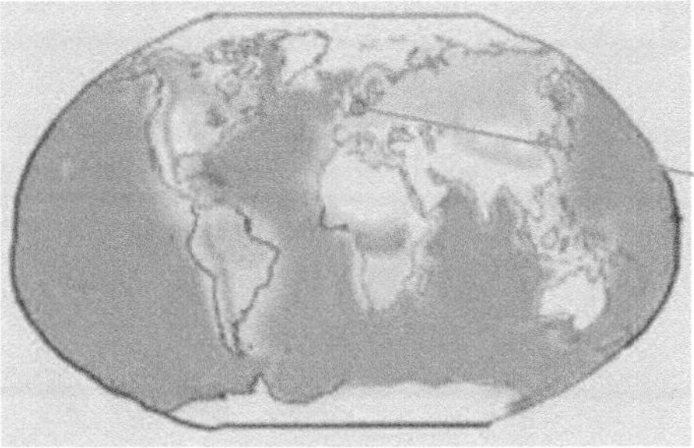

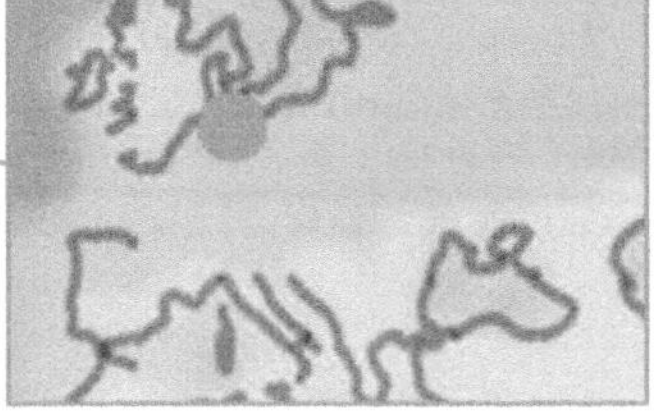

HAMBURG, GERMANY. SUMMER 1350 CE

PROTAGONIST

Agna - A bright young woman born and raised in a village on the coast of the North Sea. Her parents were fishers. Her great-great grandfather had been a soldier, an officer, and a multi-lingual diplomat in the army of Kublai Khan wounded in the Battle of Legnica; left for dead by the departing Mongol army. This secret about the family's founding is never discussed.

ANTAGONIST(S)

Matius - A Norwegian sailor, and walking friend to Agna when his ship is docked in Hamburg.

Kurt - A young fisherman from Agna's home town near Lubeck.

Meshek - A young Jewish boy that Agna meets on the way south toward the Harz Mountains following the outbreak of plague.

Sophia - A Catholic nun from Ravenna, Italy, who escaped with her Mother Superior when the plague arrived at the ports of Naples and Messina on ships bringing goods from Asiatic ports in Turkey and Lebanon. With the plague arriving at the Port of Hamburg, she must once again escape.

LESSER CHARACTERS

Friedrich - A ship's purser that Agna grew frustrated with because he disagreed with her over payment for an order of bakery goods.

Cadell - A decent man, and owner of the bakery shop near the Hamburg docks where Agna worked until the arrival of plague.

Aldrick - A old wagon driver/teamster working for Kurt hauling barrels of Herring from Lubeck to Hamburg, and salt from Luneburg back to Lubeck. Ghengis Kahn – (b. 1162 CE – d. 1227 CE) United the nomadic tribes of the Mongolian steppe and lead them along with his offspring to conquer the largest land empire in history. It extended from China in the East to Europe as far as Hungary in the west.

The progress of infection by the plague virus is this: 1.) infection from a virus-carrier by biting, or by infection from the virus directly from a pierced bubo; 2.) 7 days later: flu like symptoms develop. These include fever, headaches, vomiting; 3.) Swollen and painful lymph nodes occur in the area closest to where the bacteria entered the skin. Occasionally, these break open; 4.) diagnosis is made by finding the Yersina pestis bacteria in the blood, sputum, or fluid from lymph nodes; 5.) Without treatment, death will occur within 10 days to between 30% and 90% of those infected.

The initial symptoms of bubonic plague are flu-like, but they tend to get worse instead of better. If the buboes get into the lungs, the time between symptoms and death could be as little as a half day.

The possibility of infection factors beyond Yersina Pestis were not much known at the time of the Black Plague in Europe. It has been shown that the climate changes that came along with the little ice age beginning in

the first years of the 14 century may have been a factor in the virulence of plague in Europe when it finally arrived along both overland trade routes, and in eastern European sea ports like Messina and Genoa in 1348 CE. After a few years the effect of the little ice age had turned much productive agricultural soil in Europe away from normally sunny summers, into cold blowing rains leaving seas of cold mud, impossible to farm. Shrinking agricultural production in Europe produced large increases in malnutrition and starvation, followed by collapsing economies and social structures. Social collapse brought with it the accumulation of uncollected garbage. With the newly abundant food sources, it brought the rat – actually rattus rattus, a small but very fecund member of the rat species.

Blame for social collapse often gained focus against "outsiders." Organized and carried out by the aristocratic families with the support of the Catholic Church and the Holy Roman Empire, anti-Jewish pogroms became rampant in the decades leading up to the arrival of the plague ships.

Agna's family lived on the Baltic seacoast three day's walk north and east from the village of Lubeck. Lubeck had become the administrative center for the entire region along the coasts of the Baltic and North Seas. For over 200 years, the commercial interests along the thousand miles of coastline between Bruges in Belgium, and Gdansk up the coast 300 miles east and north of Lubeck, had been building up trading relationships. These relationships- called, informally, The Hansa - linked commercial interests in the cities of London, across the Channel from the city of Bruges in the southwest, with the City of Gdansk at the mouth of the Vistula River nearly a thousand miles to the northeast.

Agna was a pretty girl. She was a little over 5 feet tall. She was well-muscled and well-curved as a result of the years she had spent growing up in a family that earned their way by fishing and by helping the local fishermen and farmers with their gear and with the shipping and selling of their products. Fishing had not always been in Agna's family. There had been rumors that, several generations ago, a castaway had washed ashore from a ship during a storm in the Baltic Sea. The castaway was said to be Japanese. According to rumor, he had found a way to keep himself alive, and had built a small settlement that eventually became known as RabanHaven. The survivor had learned how to work and earn money helping local fishermen and had eventually married a local German girl. They had many children and grandchildren, and some of these offspring had married and had their own children in their own time.

One of these 5th generation grandchildren was Agna. Agna was slightly different from the other German children because, for those who looked closely, her eyes had a bit of an Asiatic slant to them. Her mother and father hoped that Agna's eyes would not condemn her to becoming an outsider in whatever community she chose to live in.

Agna's parents also hoped that the other part of the family history would never become generally known, even within the immediate family; The Japanese survivor who had founded their small community, was not Japanese. He was a Mongol in the Army of Obedai Kahn wounded in the battle of Legnica over one hundred years ago.

Though born and raised by a loving family in a fishing community on the Baltic coast, she is an imaginative girl, and her parents did all they could to help her learning, and, as she grew older, her desire to see more of the world. Her family had relatives in Lubeck, about three days walking southwest of their small community on the Baltic coast.

The relatives ran a small rooming house for fishermen who spent the fishing seasons working the waters off the Baltic coast. Agna was able to work to earn room and board with them. Eventually, she tired of the need to empty the pans of human excrement that went along with the lowest of the menial tasks in their hotel. She thought of other ways to earn a living, but work in Lubeck – the commercial capital of The Hansa - was limited for girls without connections.

Opportunities for work were also limited for girls without special technical skills like reading, writing, and arithmetic, or for girls without a willingness to earn money by selling sex to men with enough money to make pregnancy, or the degradation of the act itself worth the costs.

=====

Upon hearing her plan, and her resolve to leave Lubeck, the relatives of Agna sympathized with her, though they still needed somebody to empty the pans so they would not have to pay somebody else to do it. They tried to discourage Agna in her plan, but, eventually, Agna had decided that she would walk the 40 miles to Hamburg. She then resolved to work out some kind of peaceful resolution of concerns among the relatives. The relatives finally relented to some extent and gave her some money and a basket of food for her journey.

Agna left Lubeck early on a sunny day. She began walking the 40 miles along the crowded commercial road connecting Lubeck with the City of Luneburg on the Elbe, upriver from Hamburg. Many people were walking or pulling carts along the ancient, often muddy, road in both directions. Most were carrying goods for trading in their destination city: Lubeck for those traveling north, and Luneburg or Hamburg for those traveling south. Their way was complicated by the number of horse-drawn wagons carrying barrels of salted herring from the Baltic fisheries to the docks in Hamburg. The Baltic herring fishery was very productive. Some would say that the herring formed the commercial basis for the creation of The Hansa itself. The herring had to be preserved in salt. The salt came from mines in Luneburg. Whenever a herring wagon from Lubeck had to compete with a salt wagon from Luneburg, traffic on the narrow path sometimes got tangled as one tried to get past the other.

Agna knew some of the teamsters on these wagons because her family in their home village had helped them gut and salt the herring or had helped them pound the steel into the shapes necessary to make reinforcing joints and wheel straps for their wagons. Agna did not want to show her face to some of the wagon-drivers because they had not treated her or her family well in the past, and their actions toward her now that she was on her own might be even less desirable. She believed her most important possession to be the dark hood and cloak that both warmed her body, hid her face, and hid her sexuality from those who might mean her harm.

====

"Is that you Agna?" A man on a wagon passing in her direction had called her name. Agna thought she recognized the voice, but she was afraid to turn to see who it might be.

"Agna, …It's Kurt. What are you doing on this road?"

Finally, Agna looked up to acknowledge her friend from their village. She smiled but was still unsure how to answer his inquiry.

Kurt called out to his teamster to pull the oxen and wagon to the side and stop. He jumped down from the wagon and came over to greet Agna. Both of the two young people were too shy to hug or touch in public, but they did exchange smiles and warm glances with their greetings. Finally, Kurt asked, again, what Agna was doing traveling by herself along this road with its many dangers from thieves and dangerous men.

"I am traveling to Hamburg. I have decided that I want to see if I can find better work, since there is not much work in Lubeck for young girls from the villages. I left the Hotel where my relatives allowed me to work for these past several months since leaving our village. I am very happy to see you, Kurt."

"I am happy to see you as well," Kurt responded. "But the road is dangerous, and you should not travel alone. Please come with me on the wagon. I can watch out for you, and I know some people in Hamburg who may be able to help you find work and housing."

Agna climbed up on the back of the wagon and sat next to Kurt on the top of a barrel of fish. They talked about things that had happened to them since they had last seen each other in RabanHaven many months ago.

The teamster lightly cracked his whip and the ox team leaned into the wooden shoulder harness and began to move the wagon back onto the road.

Talking was hard. The road was bumpy and rutted by deep tracks in some places, and it was deep with mud in other places. After a while talking seemed like too much trouble. There was some woolen cloth in the front of the wagon. Agna retrieved it, pulled it around herself then laid down on a plank that Kurt had laid on the tops of two of the barrels for her. In spite of the rough ride, she fell immediately asleep.

Later, she woke up to find Kurt shaking her, and whispering hoarsely in her ear. "Agna. Get up. Please get up." "What is it, Kurt?" She mumbled.

"Knights. They are up ahead searching the wagons for illegal goods."

The knights were the Teutonic Knights who were sometimes contracted by the business interests of the Hansa and the Holy Roman Empire to make sure the commerce moving along this particular road is approved for transport.

"I need to know if you have papers that authorize you to be on the road. If not, I can tell them you are a family friend from RabanHaven, and you work with me in Lubeck helping me organize these shipments."

"I didn't know I needed papers to walk along the road. Will I be in trouble if I don't have papers?"

The teamsters name was Aldrick. He turned to the back of the wagon then pulled his hood aside to look at Agna. He was a much older man than Agna had thought before she saw his face. It was wrinkled, and the eyelids and the bags under his reddened eyes actually drooped. Agna recognized Aldrick as a steel smith in RabanHaven who had worked his trade for most of Agna's life and had done work with her family several times. He had seemed to be ancient over all the years that Agna knew him. His voice was raspy as he spoke.

"It will go better for you if you don't say anything, and you should try and keep your face hidden. These knights all had early family members killed or wounded by the Mongols in their last battle with them in Poland a hundred years ago. They have long memories, and they might start getting excited if they look too closely at your eyes."

Agna spoke to Kurt. "Give me some dirt and grease from the bottom of the wagon around the axle. I'll smear it on my face and try to hide my eyes. I don't think a Teutonic Knight will find a dirty little girl attractive enough to ask questions. If they do ask, I'll start to cry and look at you with a very sorrowful look. You can tell them I am a little thick. That should further discourage them from wanting to talk to me."

Kurt looked at Agna. He realized that she had learned a lot in the several months she had been away from RabanHaven.

The two knights were on horseback. They were wearing their white capes over their white singlets with the red Latin Cross emblazoned on the front. Both were armed with long swords. Two squires carried out their orders by crawling into the wagons, asking for papers, and, sometimes, demanding that a case, or carton be opened for inspection.

When they stopped at Kurt's wagon, a squire asked him a few questions. He seemed satisfied with Kurt's answers, but then asked about Agna and her relationship to him. Agna had been looking down and away from the eyes of the knights and the squires. Kurt told them his story about why Agna was with him. The story was a good one because it was true.

Agna glanced up briefly when she knew the squire was asking about her, but she quickly looked away. The squire asked no further questions. He jumped down from the wagon and walked alongside the knights to the next wagon in the line alongside the road.

Agna could not help but be impressed with the knight's dress, their carriage, and their demeanor. "They look too young to be much concerned about losing a distant relative to the Mongols so many years ago." Agna said to Kurt.

It was at that point that the teamster, Aldrick, turned to the young people, and spoke to them.

"These knights are trained from birth to believe that that they won a decisive victory at Legnica in Poland, but the truth is that the knights armies were destroyed by the Mongols. These Knights claim that their victory is the reason why the Mongols retreated from Europe a few months later and never returned. The truth is that the Mongols left Europe because they got word that their Great Kahn, Obedai, had died in their capital at Karakorum.

"Karakorum is the Mongol capital built by Genghis Kahn after he had conquered all of Asia almost to the great water beyond China. Karakorum is deep in the center of Mongolia many months ride from here. The Mongols all knew there would be a political struggle to name Obedai's successor, and they all wanted to be there to help protect the interests of their particular family faction.

"If you take no other advice from me ever again, take this: Never discuss what I have just told you with anyone. If you ever did, and the word got back to one of the family of any of the knights, that knight and that family will search the earth for you. They will never stop, and when they find you they will kill you in a very slow and painful way," Aldrick paused a moment; "...but probably not before they have tortured my name, as your source of the information, out of you."

Kurt spoke: "All the knights and their families are now part of the commercial interests of the Hansa. They are much more interested in making a few Florins off people like me who haul their salt and their fish, and empty their chamber pots, than they are in seeking revenge over old grievances," Kurt paused, "...except for those grievances that arise over any threats to the honor of their families, or their family name. Those grievances never go away."

Kurt motioned the old teamster to move out. Agna grew thoughtful. After a period of time the rocking motion of the wagon over muddy, rutted ground put her back to sleep.

====

The Holy Roman Empire had been formed under the authority of the Catholic Church for the purpose of restoring Christendom to Europe after the fall of Rome to the Barbarians 500 years before. Pope Leo III named Charlemagne Holy Roman Emperor in the year 800 CE. Among the mandates of the Holy Roman Empire, was the reclamation and security of the Christian artifacts in the Holy Lands, and the restoration of the Holy Lands themselves to the Catholic Church.

To carry out these purposes the Holy Roman Empire – under the authority of Pope Urban II in 1095 -- fostered the creation of three groups of noblemen as knights to lead Christians into battle against the Muslims who now held Christian lands and artifacts in all the holy lands. The three groups; The Teutonic Knights, The Hospitallers, and the Knights Templar – were authorized and largely financed by the Catholic Church to conduct a series of Crusades against these Muslims.

The Teutonic Knights of the Catholic Church and the Holy Roman Empire had been engaged in a complex struggle for influence within the organization and maintenance of the program of Crusades to protect the Christian relics in the Holy Land. This struggle among Christian factions of the Catholic Church had continued across most of central and northern Europe for 150 years.

None of the Crusades were successful in recovering any of the Holy Lands for the church. When the Muslims finally defeated the Crusader's attempts to secure the Holy Lands, the Crusaders gave up on the idea. There were no more Crusades into the Holy Lands after 1291, but there were sporadic efforts to refresh the focus carried with the declaration of a Crusade, particularly in northern and western Europe.

Legnica, in Poland, 150 miles southeast of Lubeck, had been the scene of the final battle between the Mongols and the leftover military groups still riding under the now mostly discredited Crusader's Cross. In 1242, the

Mongol army won the Battle of Legnica, but they did not pursue the bedraggled knights any further into northwestern Europe. Instead, they moved south into Hungary, and shortly after that they left Europe and returned to Mongolia.

The Mongols never returned again to Europe.

Wounded Mongols who somehow survived battles, but had not been collected by their departing army, did not last long on the hostile grounds of those like the Teutonic Knights that the Mongols had so ignominiously defeated.

In their common interest in fending off the westward movement of the Mongol armies, they had come together, but had lost Legnica to the superior tactics and command of the field of battle by the Mongols.

The departure of the Mongols from Europe after Legnica had given the leftover Crusaders an out, and an opportunity to refresh lost glories and pride in what they claimed as their earlier achievements: both the church and the Holy Roman Empire could claim they had actually won the battle of Legnica. With this empty moral imperative putting wind in their sails, the knights could then hunt down the surviving Mongol soldiers and execute them without fear of terrible revenge in the future by the ravaging armies of the new Great Kahn, Mongke, grandson of Genghis, creator of the Mongol Empire, and the first Great Kahn.

In the late 13th and early 14th centuries the Mongol Armies of Genghis Kahn had conquered most of Asia. He had then retired to build his great castle at Karakorum in the middle of the high Mongolian steppe. He put his army under the leadership of his children and grandchildren to carry out his vision of world dominion for the Mongol people. The battle tactics of the Mongol Army depended on lightly-clad soldiers riding fast ponies for speed and quick movements to carry out their generally superior tactics on the battle field.

One of the Mongol tactics, feigning defeat and retreat from the battlefield, was particularly effective in drawing the knights after the retreating Mongols at a fast pace until their horses, overloaded with armor, wore out. The main body of Mongols, then, charged down from the hills and wiped out the tired knights. their clashes with the Knights in their heavy, head-to-foot armor and riding horses, also armored, were too tired to carry all the weight over more than short distances. These tactics worked best when the Mongols had open fields of battle. In the surrounding forests, some argued that the knights would have an advantage.

Even chain mail on the lower ranking soldiers in the knight's ground army was not very useful in stopping the arrows shot by the Mongols. The Mongol arrows were a source of pride for the Mongol archers. Flanged, constantly sharpened, and often dipped in excrement, the arrows pierced the mail, then the skin that tore when the arrow was pulled out. The tearing caused infection that usually led to death. The mounted Mongol horsemen, on the other hand, wore body silk that wrapped itself around an enemy's arrowhead if it had buried itself in the body.

The arrow could then be pulled out without tearing the skin, thus reducing the risk of the infection that would likely cause death.

=====

On first arriving in Hamburg, Agna spent several days wandering the streets, trying to talk to people who might help her find food and shelter as her small store of money and food was now gone. In addition to imagination, Agna also has some skills, and a tolerance for hard work over long hours. She had long ago lost any pride she might have had about dirty jobs during her time growing up with her fishing family, and at the Hotel in Lubeck. It was not more than a few days before she found a place where she could earn some money, and begin, then, to think further about what she really wanted to do with her newfound freedom.

She knew how to do everything involved with deciding which herring of a batch of fresh caught herring were worth saving, and which should be thrown away. She knew how to gut the fish that were worth saving, and she know how to separate and salt all the premium parts and the leftover parts of the fish that had value.

She had her own gutting knife and sharpening stone. Her father had given them to her on her 6th birthday. They were precious to her, and she was resolved to not lose them. Agna had other skills of the kind needed around boats and the people who worked on boats. She had skills with woodworking, and with the repair of fishing nets and sails. She knew how to raise a sail, and she know how to take it down to take advantage of shifts in the wind, particularly in stormy weather in the shallow waters near the rough coasts of the Baltic.

Another precious skill given to Agna by her mother and her father was a basic ability to read, write, and count. Reading was a rare skill anywhere in northern Germany, and it was rarer still among girls and women in families of serfs normally bound for life to the owners of lands they and their families worked day and night for the benefit of the land owners. Occasionally, though, a more enlightened land-owning family might allow the more talented children among their serfs to earn a living by other means, but it was a rare thing in these troubled times.

There were many itinerants in Hamburg during the social disruptions that followed the miserably cold winters, the collapse of agriculture, and the end of the Crusades in the early years of the 14th century. The young men and women had grown up as serfs working the lands surrounding the castles of one of the local noblemen. They were not trained in any useful skills outside the heavy work – mostly lifting and shoveling - of farming, animal husbandry, and house-cleaning.

====

With Kurt's help Agna had found work as a baker's helper in a shop that processed and sold flour and bread to pursers on ships traveling the Elbe River between Hamburg and the North Sea coast. When those ships needed to load or unload cargo, then re-stock, they tied up at the Hamburg docks near the shop where Agna worked. The shop valued Agna's skills as a baker's helper, and they valued her ability to speak more languages than the crude German of most of the young German itinerants looking for work every day on the docks.

One sailor, Matius, came into the Port of Hamburg every few months during the past year. One day, he had come into the bakery to buy a loaf of bread and had met Agna for the first time. They had grown close. When Matius was in port, he and Agna would walk around the City, and enjoy the sights, sounds, and smells together.

Matius told her that he and his fellow sailors all seemed to dwell on the stories of those who had caught the plague, and then died a horrible death from it, sometimes within a day or two of getting sick. The captains of the ships, on first hearing even a rumor of plague in a port, would immediately untie from the dock and head to sea; whatever the weather; whether or not all the sailors were aboard; and whether or not the loading and unloading, had been completed.

"Whenever you hear the word plague, you must immediately gather some food and water, and leave the city," Matius said to Agna. "Evan an approach to the man or woman who has used the word to find out more information, may kill you. You must promise me that you will leave immediately." Matius looked deeply into Agna's eyes.

"Yes, Matius, I will." She replied.

====

Once when there were several boats tied up at the docks, and the baker's ovens were working constantly, the bakery owner usually sent Agna to the boats to ask the purser about supplies of bread and flour they might need for coming journeys. When not actually selling, she often accompanied the bakery's order to the boat to make sure the purser got satisfaction with the order. When unsatisfied with the order, and uncompromising with Agna, she had learned when to demand that the purser take her either to the Captain, or to the boat's owners if they lived in Hamburg.

Agna quickly learned the combination of charm, sexual innuendo, more assertive talking, yelling and foot stomping needed to resolve these kinds of commercial problems.

Once, when she grew too frustrated dealing with a ship's purser named Friedrich, she went to the rooming house where she knew Matius was staying. They had not known each other well, yet, but she trusted Matius to do right by her, and he had the size and demeanor of a man who could be ready for a fight when necessary. Even better, Matius knew how to look the part of a very angry man.

On that day, Agna knew that the bakery owner was away, dealing with a shipment of grain needed by his shop. He was not available for counsel about how to deal with the problem purser. It was all up to her.

Matius readily agreed to come with Agna to the docks. They arrived to find the purser back at his desk hunched over his paperwork. Matius motioned Agna to go back down the gangway to wait for him on the dock. Matius then barged back into the room without knocking. The purser looked up, his face suddenly pale. "What do you want?" He asked.

"I want the money you owe my friend for the bread and flour she delivered this morning," Matius said.

The purser replied: "The bread was stale and the flour riven with weevils."

"Show me the flour," Matius said. "Now."

"No." was the reply."

Matius picked the purser up by the collar of his shirt, carried him to the door, and through it to the rail outside. Before the purser could grasp what was going on, Matius tipped him over the rail and dumped him into the water 20 feet below. Matius took a moment to make sure the purser could swim, then turned to go to find the ship's cook.

The cook had been watching as the purser went into the water. He moved to intercept Matius, then took him to the Captain who was passed out in his room from drinking too much beer. The cook shook the Captain, hard, then kicked him. "Captain! Wake up! This man wants to get paid, and the purser is acting like an asshole again."

The Captain muttered something about the keys being in the desk drawer. Matius thought he heard the Captain mumble something to the effect of "Pay the bastard, and then get him the fuck off my boat."

The cook looked at Matius and winked. "How much?" The cook asked.

"Four Florins." Matius replied.

The cook used the Captain's key to open a drawer. He pulled out a small leather bag and shook out the four Florins. "No hard feelings, I hope. There was nothing wrong with the bread or the flour. Our purser likes to think he is a tough guy who can bully people, especially pretty girls like your girlfriend, so he can cheat them out of their money. Then he takes the money that should have gone to the supplier and stuffs it into his own pocket."

Matius, not knowing how to deal with this strange crew, nodded, but did not change his expression.

"This garbage scow may want to return to these docks again. If I am still aboard, I hope we can do some more business. Once I talk about this incident with the ship's owners, I doubt if you will see either the Captain or the purser again. Can we shake on it?" The cook extended his hand.

"Yes. There are no hard feelings. I'll let my friend know what you have told me, and that you have assured me that she and her baking company will be welcome to supply your ship in the future."

Agna looked up to Matius as he came down the gangway. She had grown concerned when he threw the Purser overboard, then disappeared inside the ship with the cook. She hugged him.

"Here's your four Florins," Matius said as they walked away from the dock.

====

Hamburg is located on the Elbe River about 40 miles inland from the North Sea coast. Since the formation of the commercial Hansa in the early part of the century, Hamburg has become an Important shipping port built up to handle shipping between the farms and produce of northwest Germany, and the European markets scattered around the North Sea, the Baltic Sea, the channel between England and France, and the distant ports on the far seas beyond.

Agna knew that plague might come into her city of Hamburg. As the rumors of plague in other European cities grew more numerous, she grew more concerned. She heard more stories as some of the shopkeepers had begun whispering in the nearby market. She knew she would have to get away from the docks, soon even get away from Hamburg itself, if she was to survive this plague. Agna did not know exactly what a plague meant, but she had heard some of the older people in the dock area, and some of the sailors who took a few days leave from their ships talk about it.

====

When Agna got word that the plague had come to the Hamburg docks, she gathered some clothes, and what money and food she had on hand, and began to walk south toward the Harz mountains, about 100 miles away. In the mountains, she thinks, the cooler air might stop, or at least slow down, the spread of the plague before it could catch up to her. She had heard rumors that this would be true from stories told by travelers who had escaped plague in England, Ireland, and Scotland the year before.

She is not alone on the road. Others are walking, some from the upper classes ride in horse-drawn carriages. Agna is dressed in her flour-stained work clothes from the Bakery. Several others on the road are similarly dressed, as though they felt they had to escape the city as quickly as possible, grabbing only the few hands full of personal items as they ran out of their homes and into the streets; ...as indeed they had.

As the day unfolds, more people join those walking south. Agna is strong from her upbringing on the Baltic coast, and from lifting and carrying the bags of flour at the bakery. She resolves to walk as far as she can on this first day. When the time comes, she thinks, she will go into the woods and try to sleep a little. She knows, though, that the road will only grow more crowded with refugees from the plague, and she has no idea what might await her in the mountains. She knows she can only sleep for a little while.

Some of the wealthier residents of Hamburg are moving along the road in horse-drawn wagons. They shout at those ahead of them to get out of the way, as though their passage was more important than the struggles of the walkers carrying all their belongings on their backs. Agna resented this, but had no will to shout back, or express her resentment in any other way. Ahead of her, a figure in a hooded cloak sat huddled alongside the road and hugging their upper body as though in pain. It appeared to Agna that he, or she, might have been struck by a rock or piece of wood thrown up by one of these fast-moving wagons. As she approached the huddled person, a girl, looked up, crying. Agna notes that she wears a white hood inside her outer cloak. She rubs her side below her left breast. She asks for help in a foreign accented German. Agna sees a piece of wood as large as her leg laying in the road nearby.

She lifts the hooded girl to her feet and struggles to help her move further off the road and behind some bushes, so she can look at the wound. Agna pulls the cloak and her inner garments, both ripped where the wood had struck the girl, aside to look at the wound. She sees a black and blue patch of skin, but no blood. She had nothing to help the pain, but the wounded girl seemed to calm down as Agna ran her fingers over the wounded area of her body, and then laid her palm on the black and blue mark.

The girl seemed to calm down and take comfort from Agna's healing touch.

In a careful German, Agna asks how the girl should be called.

"My name is Sophia. I am a nun with the Catholic Church in Lubeck." Sophia looks carefully at Agna. She searches Agna's face for any reaction to her words, but she saw none. "Four days ago my Mother Superior told me to take a few of my most precious belongings and some food, and walk as fast as I can to the Hartz Mountains. She said that a plague ship will arrive in Hamburg in 5 days. I became very frightened and left right away."

"I think that ship will arrive in Hamburg sometime today, unless the city leaders can persuade whoever is sailing the ship to stay away. I don't think that will stop the plague, but each day they are stopped from docking, is another day the plague can't begin to move through the City." Agna said. "Can you walk?" she asked.

"Yes. But not as fast as I would like. Also, these pages of parchment and paper are heavy and unwieldy." Sophia opened her cloak further to show a pouch she had tied to her body on the left side. The pouch had probably stopped the wood staub from puncturing the skin covering her ribs. In the pouch were some pages with writing on them.

"What is written on these pages?" Agna asked.

"These are words written by a church scribe in my monastery in Italy; …in Florence. His name is Teodoro. He was a scribe to Boccaccio in Ravenna. He copied some of the stories Boccaccio started to write about the plague. Do you know Boccaccio?" She asked. Agna shook her head, no.

"Two years ago when the plague first came into the ports of Naples and Messina, there was a lot of panic all over Italy, among all the classes. Boccaccio wanted to capture the thoughts of some noblemen and women of that area as they tried to think how to escape the plague."

Agna looked at one of the scripts but could not make any sense of it. It was written in Italian, first of all. "I can't read these words," she said. "Can you read them for me?"

"I will try to translate them," Sophia said, "but you may have to help me find the correct German word."

The two women went on in that way for another couple of miles. By that time, the sun was getting close to the horizon. "We will need to stop before it gets too dark. We need to find a safe place to sleep. We are both tired, and we need to rest. How do your ribs feel?" Agna asks.

"It hurts terribly, but I can walk with it." Sophia replied.

Agna could see the grimace in her face as the nun spoke. She knew they would have to stop soon. She also knew that two young women on the road without any visible escorts were an invitation to trouble. Agna knew herself to be strong enough and smart enough to discourage most fainthearted attackers, but she was not sure what to do if she had to defend the both of them against an attack by one who is dedicated, smart, and hungry for whatever two women might have to offer.

She slowed down to allow Sophia to take a slower, less painful pace. She looked around the shallow hills on either side of the road in the hopes of finding some shelter. Two mature ravens flew over them toward some trees to the right. One of the ravens seemed to turn its head to look at Agna, but then turned back to continue on the way with its mate.

At the sight of the ravens, she felt a twinge of nostalgia for her home in RabanHaven near the shores of the Baltic. RabanHaven means home of the Ravens in German.

Agna noted that the birds were flying toward a small copse of trees to the west that could at least provide shelter from the wind. The two young women had been moving fairly fast on the road, and the crowd on the road had thinned considerably as night closed in.

Agna kept walking to let a little more light fade from the sky. Then, suddenly, she grabbed Sophia's arm and pushed her off the road, across a ditch and up a small hill on the other side. They then crouched down to walk furtively toward the group of trees where they hoped to find some shelter for the night. Agna hoped that no one had seen them leave the road, but it sounded like someone had.

She whispered to Sophia to stop for a moment and sit quietly. Sophia did as she was told. Agna turned to observe the other figure so she could decide whether the other, who had also left the road in the same place, might be a threat to them.

Suddenly, the other called out to them in an accented German from about 20 paces away: "Hello. I don't mean you any harm. I only hope to find someone to travel with. I think we will do better sharing what we know, and whatever else we have as we try to get to the mountains. I have some food, and a goatskin of water."

Agna whispered hoarsely, "Who are you?"

"I am Meshek. I take care of the accounts for one of the members of the Hansa who sells saddles and harness leathers in Hamburg. I hope the plague passes by without too much harm so I can go back to work there." "What will you do if the plague lingers?" Agna asks.

"Then I think we are all in trouble," Meshek said. "But even that terrible eventuality might still be a good reason to share our resources. May I ask your names?" Meshek asked.

"Yes. I am Agna, and my friend is Sophia. We are both strong and well-armed with knives and clubs. How do we make sure you are no threat to us?"

"What kind of assurance do you need?" asks Meshek. "The only real threat any of us has is the threat of infection from the plague. I'd say that, if we have made it this far from Hamburg today, and we are each still in good health, then we are probably not a threat to each other from plague. Now you say you are well armed and strong. I am only armed with food and water. So you have me at a disadvantage. Are you hungry?"

"Yes, we have some food, but not much. Do you have enough to share?" Agna asks.

"All of our food will need to be carefully rationed, and I am willing to put my whole stock together with yours so we can share equally. You say you have knives and clubs, and the strength to use them. You should put yourselves at ease, then. Though I am a rather tall man, I have no weapons at all, and have never been in a fight in my life. I keep books after all."

Sophia spoke. "You sound like a good and careful man, Meshek. I think, Agna, that we can trust him."

"Yes. Alright," Agna said. "But please walk ahead of us and a little distance away so we can get used to traveling with you. We are headed toward that small copse of trees up there. I want to get there before it is totally dark so we can see if it will work as shelter for the three of us."

The three travelers made their way to the trees, arriving there in time to look around. The place seemed suitable, so each moved some branches around to give themselves some comfort. Agna shared some of her bread with the others, and Meshek passed his goat skin of water around to each of them.

Two ravens came around to see if there might be a handout. Agna was certain that they were the same two ravens that had guided her toward these trees. She tossed a small chunk of bread to each of them.

Agna noticed that one of the ravens had a single white feather on the top of its right wing.

"Your accent is strange to my ear, Meshek. Where are you from?" Sophia asks.

"I am Jewish," Meshek said. "I am from many places. And you are Catholic. I can tell from your white coif, and what is left of the black in that part of your robe not covered in mud. What is your order?"

"I am Benedictine. My Mother Superior and I left our church in Ravenna, Italy to escape the plague there, but now it has caught up to us again. By an arrangement with the Bishop here, we have been working at the Cathedral of Saint Mary in Hamburg. On hearing of the plague ships, my Mother Superior felt she must stay in Hamburg to help the sick there, but she told me that if I did not leave immediately for the Harz Mountains she would make sure that I would be thrown out of the Benedictine order, and never allowed to return.

"There is a Cathedral at the foothills to the Harz Mountains called the Walkenried Abbey. Mother Superior has given me a note to the Cistercian monastery there, to ask them to take me in."

"Do you think they might take in a Jewish man?" asked Meshek.

Sophia answered. "The Cistercian Order, and my Order, the Benectines are not allowed to turn travelers away. We must give them shelter and food, and we must bind their wounds until they are well enough to be on their way. You should have no fear of them."

With that, all three fell into a deep sleep.

====

Agna woke up hearing noises along the road. In the east the night sky along on the horizon was turning grey with approaching sunlight. She roused her travel mates. "We should be on our way while the road is still mostly uncrowded. Here is a bit of bread to chew on."

Meshek passed the goatskin of water around and contributed a bite of rabbit meat to the common table. "We will need to find some food soon," He said. "I have a sling and a knife, and I am a pretty good shot with the sling. If you see any rabbits, or any abandoned animals of any kind except the ones that look sick, let me know. If they are nearby, I may be able to cripple or kill it with my sling."

I think we need to look for a way toward the mountains that will be less crowded, and less dangerous than this road. I fear that recent escapees from Hamburg may have been exposed to plague, and they may infect one of us," Agna said. "How are you feeling, Sophia?"

Sophia had her body turned away from the other two. Her head was bowed as she made her morning prayers. She remained silent for a few moments, then turned toward Agna and Meshek.

"My side is still painful, but I think I can keep up with you through the day. Do you want to go now?" She asked Agna.

"I think we need to start soon, but I'd like to talk a bit more about finding an alternative to walking on the road."

Meshek spoke. "I think we will make much better time on this road. For one thing, we know where this road goes, and we might find others, like ourselves, who have food and water, and would prefer to travel in a group." He continued. "If we go across the fields that I see here, we will spend a lot of time in mud, and thick brush. Also, if we find a road, we won't know where it goes unless there are already refugees upon it. If there are refugees on any new road that we find, then we will have wasted the time we could have spent walking on the road we already know, and the progress we could have made toward higher elevations."

"You make compelling arguments, but I am still concerned that this road is the main road to the mountains, and it will become very crowded as the day goes on. I would have no doubt about looking for another route if it were just myself and Sophia, but your presence gives me some hope that threats against she and I will be less while you are here.

Agna continued. "But you have told us that you are unarmed. You have suggested that you are a coward when presented with the possibility of physical conflict. And yet, this morning you have told me that you are armed with a sling and a knife. So, please help me understand. Will you shrink from a fight and run away at the threat of violence, or will you stand?"

Meshek thought for a moment, then rose. "I will answer your question, Agna, but we should get moving, and I propose that we stay on this road for now. As I said last night, the most important thing to me is traveling in a group. I am prepared to give whatever I have to any group that I believe I am compatible with, as I believe you both are."

"I do have more to contribute to such a group than what I let on last night. I am skilled with personal combat using knives and pugil sticks, or whatever heavy, blunt instrument is at hand. I am skilled in the martial arts that I learned at the hand of masters of these arts living in the Asian precincts of London."

Sophia was listening intently to this conversation. There were tears in her eyes. This morning, her native Italian dialect more deeply colored everything she wanted to say to the two German speakers. Her voice quavered as she spoke. "I can't help in any of this. I left Hamburg with almost no food or water, only whatever scraps my Mother Superior was able to give me as she brought me to the road before us now. I am completely dependent on you both to help me. ...especially now that I am in pain. Please stay with me, please help me."

Meshek looked toward Sophia for a moment, then back toward Agna. He said nothing more, but both Agna and Sophia rose to begin walking with him back to the road. Meshek stayed to the side and a little ahead of the two women. Meshek believed his presence would be more valuable to the two women if he was not seen to be directly associated with them.

As they regained the road, the sun was fully above the horizon. Agna expected the day to be hot. Meshek seemed to have water, still, in his goatskin for today, and she had a little as well. But they would certainly need to find water tomorrow. Agna believed that they would need another three days after today to get to the edge of the woods that marked the foothills of the Harz mountains.

As they walked, Sophia had told Agna of Walkenried Abby in the foothills of the Harz Mountains. It is an Abby of Cistercian monks and nuns. The Cistercians were associated with the Benedictines and they shared the values of hard work, particularly in agriculture, and particularly in their home communities. They ran mines in the mountains and they used the proceeds of their labors to help finance some of their commercial activities with the Hansa in Lubeck and Hamburg. It was rumored that they even used some of their funds to aid some of the activities of Pope Clement VI who conducted his Papacy in the French city of Avignon in southern France northwest of Marseille. In that way, the church would leave them alone to work out their necessary arrangements with the Hansa.

The Cistercians had helped to finance some of the Crusades, and some of the Crusader knights financed by them had fought in the Battle of Legnica over 100 years before. The Cistercians, and, through them, the Benedictines, including nuns like Sophia, were generally held in good standing wherever they went in northern Germany.

"I am offering these insights to you," Sophia said, "… so that you will see some value in my presence with you."

"Please don't worry about that, Sophia. I am happy to help you on this journey that is so important for all of us. I know you would do the same for me, and probably even for Meshek. You must not think that either of us will try to find you some day after this is all over to lay claim against you for services rendered during this terrible plague. I can't even imagine such a thing in the ruins of society that will follow the end of the plague."

Meshek had been walking behind, but now caught up with the women. "I think we should stop for a bit to rest and eat something. Let's move off the road away so we can have some privacy. I think I saw another rabbit. If I can catch it, I won't want to share with anyone but the two of you. "

The three moved to a small mound of earth in a cluster of small trees about 50 paces off the road. Meshek whispered that he would return shortly. Sophie expressed her relief to Agna by smiling slightly in her direction through her pain and fatigue.

The road had grown more crowded. There were more walkers carrying their bundles. They looked tired and downtrodden, especially when the wagons of the wealthier commercial classes passed by, kicking up rocks and the layers of dust covering the road, a dry powdery dust drained of moisture by the hot sun.

Shortly, Meshek returned with a dead rabbit. "I brought this little fellow down with my sling. He didn't go willingly, but I'm sure he will be grateful to have given us needed sustenance on our journey."

"How good are you with that sling?" Sophia asks. Are you like David taking on Goliath in the valley of Elah?"

"Well, we Jewish people don't like to take on airs, so I will defer any comparisons with that David. However, I am pretty good with the sling, and I have had to take down the occasional hostile threat with it."

"I think you need to tell us more about who you are." Agna said. "I believe you are a good man, but I also believe you have a troubled, or perhaps a troubling, past. Do you, …have such a past, I mean?"

"I have a past, as we all do. As a Jewish man who believes in his religion, though perhaps not in the rigorous practice of it, I have been many places, and done many things to survive in a world, especially a Christian world that is largely hostile to me. I Have learned what I know of the sling from the story of King David, and I have learned what I need to know of combat without weapons from the Asiatics, who, unlike Christians, neither fear us, nor seek to destroy us as a people.

"I could have made myself very much at home with the Mongols, if I had lived in those times, and if I had not found a good position with some business men in Hamburg who valued my skills and talent more than they feared my Jewishness. Actually, my final skill, with knives and other piercing instruments, I learned from these businessmen. Their specialty in Hamburg's commercial world is the cutting and stitching of heavy leathers used in saddles and harnesses. Having these leather-working tools, and knowing how to keep them sharp, clean, and ready for use is a matter of strong pride with me."

"You have many reasons to fear my skills in the violent arts, but, I think, many more reasons to view me as your protector, as that is my intent with you, and with Sophia. I know Sophia is a member of a Christian order, though not one with violent intent toward the Jews."

"That is true," Sophia said. "We, the Benedictines and the Franciscans, are more interested in helping things to grow into healthy adults in maturity, than we are in seeking revenge against ancient enemies as a way to build congregations of loyal adherents to their violent dogmas as many Christian sects are."

Meshek nodded in Sophia's direction. "Perhaps we should be moving along."

"Not yet," Sophia said. "I want to say something. I do have something to help us in our journey besides a possible place to go to in the mountains. I've shown this to Agna, Meshek, but I want to show it to you as well.

"When Mother Superior and I decided to leave Ravenna to escape the plague two years ago, we decided to take some of our personal artwork as bargaining material for food, or shelter, or whatever we might need as we made our way overland to northern Germany. Our Order in Ravenna had a very good reputation among the arts community in that region, and artists who had received favors from us would often repay us with their works. Some of those gifts went to the church, but some went to Mother Superior and I, because of our work in the fields for our order. Our charitable works outside the walls were also very noteworthy because of the number of people who were very grateful for the service we had rendered to them. They often offered gifts to the church in our names in recognition.

"The poetry of Dante writing his 'Divine Commedia' in spoken Italian early in the 13th century inspired many of the artists writing and painting in the later years of that century. Since the beginning of the 14th century, the beautiful paintings of Dante's contemporary, Giotto, had greatly influenced the art works not only of the later visual artists around Florence and northern Italy, but even the writings of our good friend in Ravenna, Giovanni Boccaccio. To some extent, Boccaccio's contemporary, and older friend, Petrarch was also influenced by Dante, though Petrarch was a refined man who expressed discomfort with the vigor of Dante's writings in vulgar Italian.

"As we were getting ready to depart Ravenna, rumors of plague were everywhere. Though we were making haste to leave, Boccaccio seemed to relish all the rumors of the plague's grisly effects, and the feeble efforts of the wealthy to not succumb to an illness of the poor. He began to write some stories, and he gave me a draft of one of those on paper. Even the paper itself was expensive, but nothing compared to the value of anything written by Boccaccio.

"We had pieces of work from all of them. I even had a small sketch done in ink on parchment by Giotto and given to his patron in those early days. His patron later came on hard times. Our Order gave him food and a place to stay, and he repaid us with Giotto's beautiful sketch. My Mother Superior insisted that I take the sketch for safe-keeping as we traveled north from Ravenna.

"When it was time for me to leave Hamburg, my Mother Superior would not hear of me giving the sketch back to her. Tears of pain and longing came to her eyes but she would not accept the sketch from me. She turned her back on me, and it was then that she threatened to have me thrown out of the Order, something I could not bear.

"I still have the sketch with me. I have been able to keep it with me for these two years, but I fear the loss of it now that the plague stalks us, as it stalks the desperate people who walk beside us along this terrible road.

"Everything in this bag is a treasure worth, altogether, an unimaginable value in the right time and place, but for now, I would almost trade all of it for safe passage to the door of Walkenried Abby." Sophia's head drooped now that she had exhausted herself of all she wanted to say.

Agna put her hand on Sophia's shoulder then pulled her toward her body in a warm hug. Meshek reached out with his hands to touch each one on the shoulder.

"Now we must go," Meshek said.

The three travelers rose and re-joined the walkers on the road in their continuing walk south, away from the black death creeping along behind them.

=====

Once again, on the third day of their journey, Agna rose first as the sun was only just beginning to turn the horizon in the east slightly gray. This morning, though, the gray seemed to have a reddish tinge – a sign of almost certain rain.

Agna shook the other two awake. "Get up. We need to make as much progress as we can before the rain gets too close to us. We will need to find a shelter from it, or we will need to make a shelter from whatever we can find in the woods."

Fortunately for the travelers their increasing elevation had produced a few more trees along their route. The trees could provide shelter and warmth; even the possibility of having some rabbit meat cooked over a fire. The idea made Agna's mouth water. The other two were still too groggy from waking up to care.

The rain came down heavily for an hour or two in the morning. Then, the clouds cleared, the sun came out to once again dry the roads. By the afternoon, the winds began to kick up the dust from the roads, and it began to seem that each particle was destined to get into the eyes and nose of each traveler.

Soon the three travelers resolved to find a dry place out of the wind, so they could spend a day finding some meat and some firewood to cook it with. If they could find a stream coming down the mountains they could even consider taking their clothes off so they could clean their bodies. Once they found just such a place, they could not help but feel overwhelmed with their good fortune in having survived this far and having each met and made friends with two good and supportive people.

Once the three had their day of rest, their good fortune continued when Meshek hailed a fellow driving a wagon pulled by two oxen. In the back of the wagon were two barrels of salted herring for delivery to the Abbey. Meshek recognized the harness on the oxen from having been sold by the company he worked for. When he pointed this out to the teamster, the teamster invited the three of them to ride along with him for the remainder of the journey to the Abbey.

On arriving at the door of the beautiful Abbey, Sophia's letter was gladly accepted, as were the two barrels of salted herring, and all were invited to join the Cistercians at their evening meal.

Before entering the building, Agna looked around the surrounding woods in the hopes she would see some kind of vision that would help her decide what to do next. Sophia was walking next to a Monk of the Abby. She turned and looked back at Agna and smiled.

Meshek also looked around at the surrounding woods, sky, and the Abby grounds. As his eye caught the eye of Agna, he, too, smiled.

PANEL 6 - REBIRTH

Orabella, a homeless young woman, now apprenticed to the Studio of Verrocchio, paints a fresco for the home of Lorenzo de Medici in the hills above the Tuscan City of Florence. She has the help of the young Leonardo da Vinci in the design of the work.

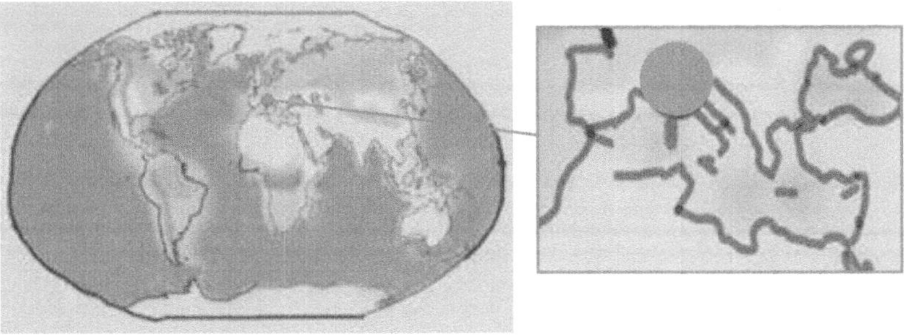

FLORENCE, ITALY. SUMMER 1475 CE

PROTAGONIST

Orabella - A pretty and bright young orphan girl, taken from the streets, and into the household of Lorenzo de Medici as a maid and cook's helper. Through talent and hard work she quickly rises to become one of the household supervisors in charge of the kitchen and cleaning staff.

ANTAGONIST(S)

Leonardo da Vinci - (b. April 1452 – d. May 1519) a young visual artist and polymath – 23 years old in 1475

Andrea del Verrocchio (b. 1435 – d. 1488) Age 40 in 1475. A master of perhaps the most prestigious of artist workshops in the City of Florence Italy.

Lorenzo de Medici (b. January 1449 – d. April 1492). Age 26 in 1475. Head of the Medici household; married to Clarice Orsini in February, 1469. He had 10 children, one of whom, Giovanni di Lorenzo de' Medici, became Pope Leo X in March 1513. A major patron of Verrocchio's studio.

Adolfo Antonio - A personal messenger and agent of Lorenzo Medici.

Antonio Sforza - another young apprentice in the visual arts with not much talent also apprenticed to Verrocchio, and through him to Amadeo's supervision in the fresco studio.

LESSER CHARACTERS

Various other apprentices and journeyman supervisors who work in Verrocchio's studio in Florence.

Amadeo - Manager of the minerals used in paints, and the materials used to make the plaster for frescos. A specialist in the preparation of plaster to be used in Frescos.

Maso - A painter on wet Fresco specializing in darker blues.

Vincenzo - A mixer and grinder of pigments, with experience working with Lapis Lazuli from Afghanistan. He grows frustrated trying to grind Lapis and mix media into it in a way that makes it keep the bright blue and gold colors over the life of a fresco.

MINOR CHARACTERS

Anselmo - A butcher in the shop on the Ponte Vecchio; Alfredo - A rock smith who works for Matteo; Matteo - The owner of a rock wholesale yard on the docks west of the Ponte Vecchio.

Vitale - An apprentice in Fresco. Helps to Build frames and prepare walls for frescos; Poldi - An apprentice in fresco. Helps to Mix and apply plaster; Fedele - An apprentice in fresco. Helps organize paints; Fons - Helps to fire the kiln for making calcium carbonate out of raw limestone; Elia - An apprentice in fresco specializing in creating the cartoon from the artist's design to be applied to the wet plaster.

The Black Plague in 14th century Europe killed millions of people. So terrible were the deaths, and so devastating the effects of the Plague that the institutions of the Catholic church could no longer lay claim to the unquestioning faith of the people. Adherence to church teaching could no longer be the only sure path to happiness, or to enlightenment. Freed from Catholic dogma, however, new ideas and new approaches to old problems were beginning to take root. Guttenberg's invention of the printing press in 1450 made possible the wide distribution of new ideas on cheap paper. As new ideas spread widely, they created more turmoil and angst across Europe. Inevitably, knowledge of alternatives to the word of God were more widely discussed and debated.

Out of this turmoil, the Tuscan City of Florence in northcentral Italy became a hot bed of innovation in politics, the arts, and governance.

One great Florentine family, the Medici, not only created a stable platform to support the creation of new forms of artistic expression, they also hired many of the City's residents and supplied bank financing for their development projects. The Medici also named four Popes of the Catholic Church.

With their many commissions for the creation of works of art, architecture, and many works of great beauty, they became patrons and financiers to an artistic master and studio owner like Verrocchio, who, in his turn, provided a stable platform for the training of youthful artists, brilliant with promise like the painter and designer, Leonardo De Vinci, and the great sculptor, Michelangelo.

Even more, the Medici, under the leadership of the head of family in the mid 15th century, Lorenzo de Medici, endowed museums like the Uffizi for the display of completed objects by these, and many other of the brilliant artists of this day and age.

Florentine artist and studio master, Andrea Del Verrocchio, had been working with a group of his apprentices on a commissioned work for Lorenzo de Medici.

The work, a large fresco to be installed on a large wall space in one of Lorenzo's estate houses in the hills north of Florence, had proven to be very troublesome, even in this early design stage.

As master of the workshop, Verrocchio, in the middle of his attempt to, once again, mediate the heated disagreements among the apprentices on how to proceed with the design work, was interrupted by a messenger from the client. The messenger, Adolfo Antonio, had been observing the discussion from a corner of the workshop. He motioned to the Master to join him in a private conversation away from the apprentices.

"My master, Lorenzo, has become aware of the problems you are having with the design work on this project. He wishes to help in any way that he can."

Verrocchio responded: "I believe our problems and disagreements are not out of the ordinary with a project of this importance, but…" Verrocchio paused a moment as he noticed that the messenger had something specific in mind. "Please. I very much want to hear of any suggestions your Master may wish to make. to me, or any requests he may wish to make of me."

Adolfo spoke. "The Master has a young woman working as an assistant cook and maid in his household. In the mind of my Master, her real talents are greatly wasted chopping onions, mashing potatoes, and scrubbing floors."

"I can't imagine your Master even taking note of a household maid, unless she is a truly extraordinary person. What are these talents that can draw the attention of such a powerful and honorable man?" Verrocchio inquired.

"Once she was cleaned up and given some clothes and basic manners, it turns out she is a very pretty girl, and very intelligent. She came to us as an orphaned waif found on the streets by one of our friends who thought our family could help foster what seemed to the friend to be innate talent and character, and, thus, become a more useful citizen." Adolfo replied.

"Why does your master think she would be any help to me? If she is as pretty as you say, she will, for that reason alone, cause far more problems in the workshop than she will be able to resolve no matter how talented she is. I have never allowed women into these studios no matter what their skills are. Among these head-strong male apprentices - who are often more interested in each other than any woman can hope for - such a woman will either distract them from their work, or she will trigger the kind of male competition that can result in violence and destruction of my property. Has your Master given thought to any of these things? Verrocchio did not wait for Adolfo to answer.

"In any case, I cannot see how her presence can help resolve our artistic problems in the way of completing this fresco to your Master's satisfaction. How long has she been working in your Master's household?"

Adolfo replied. "She has been with us for over a year. She learns very quickly. She began by mopping floors and peeling onions in the kitchen. She progressed rapidly to making desserts, even designing the decorations on cakes. Her creations in the kitchen have been prized by my Master. He has said he cannot get enough of them.

"Even more importantly, she progressed into a supervisory position among the cooking staff without losing their support for her creative work. She had as much success with the household staff. She was able to dissolve some tensions that had built up over a period of months to everyone's satisfaction."

Verrocchio replied. "You have given me reasons to think more about your Master's suggestion, but I have one last question. Why would your Master be willing to part with someone who is such an obvious asset to his household. Is there some problem you are not telling me about?" He asked.

"Whatever you may think of the motivations or intent of my Master, you should know by now that he understands the careful development of natural talent, particularly if those talents emerge while they are members of his household. He recognizes the potential of this young girl, and he also recognizes that supervising a household will not hold her attention and her abundant imagination for long. He believes that the next step in her development is in your workshop. However, please understand that this is not a formal request from my Master that you take her into your workshop.

"He merely asks that you consider the idea. If you truly believe that her presence will so disturb your work on his commission that it will suffer significant delays in execution he will not further pursue the idea with you. Verrocchio thought to himself for several moments. Adolfo waited patiently to hear his thinking.

"You and your Master have given me very much of interest to think about. You have presented a strong case for my acceptance of this young girl even if she might make my life more difficult than it already is. By the way, what is her name?"

Adolfo replied. "She says a kindly old couple who took her in many years ago - now passed away - gave her the name Orabella; she does not know her surname. She has told me that she prefers to wait until she finds a surname she likes before taking one to keep for the rest of her life."

"A very wise choice. Does your Master have a period of time in mind before I give my decision?"

Adolfo responds. "None that he has told me of, but, based on my overhearing your discussions with your apprentices, I should think you would want – in your own interests - to make a decision within the next couple of weeks."

Verrocchio nodded. After a pause, he said: "I want to confer with my artistic associate, Leonardo. If I decide to take her into my workshop, I will want to make sure that Leonardo is willing to share any risks of her presence, and will be at my side when facing the inevitable difficulties among my apprentices; even better, that he be willing to give Orabella access to some of his own work to help in her education and development in the arts. As she grows in her own work, she will be able to give us a better idea of her real value to the workshop. Can you return in a week?" Verrocchio asked Adolfo.

"Yes." Adolfo replied. "My compliments on the wisdom of your decision in this matter, Andrea. Until then…" He touched the brim of his hat, turned and left the studio.

Verrocchio turned at the sound of a piece of plaster in one of the test frescos breaking and crashing to the stone floor. One of the apprentices had just expressed his criticism of a design offered by a rival apprentice by throwing a hammer at it. Verrocchio did not speculate on whether the hammer-thrower might have overheard his conversation with Adolfo.

====

Verrocchio gave a lot of thought to what he would say to Leonardo about Lorenzo's suggestion. Bringing an attractive young girl into the studio on any pretext would risk considerable disruption, and possible damage both physical and in many other ways. To survive, the girl would have to be extraordinarily capable as well as very assured of whatever skills she felt she possessed. When the inevitable artistic conflicts arise between her and some of the other apprentices, she would have to be fast on her feet, and she would have to win any contest; even the purely physical ones.

Verrocchio had almost succeeded in talking himself out of the whole idea of bringing the girl into his workshop. It was just too risky, and Leonardo would never agree to take on any part of the risk anyway.

====

Verrocchio, a very successful painter, sculptor, and goldsmith, is the master of the most successful art studio and workshop in Florence. He is a successful businessman, and an astute judge of character. He has attracted the best and most talented visual artists, sculptors and craftsmen from all of northern Italy to his apprenticeship program. Leonardo, son of Piero, born 22 miles west of Florence in Vinci, is one of these. As young as Leonardo had been when he first came to the workshop at the age of 14 in 1466, Verrocchio recognized his genius early. Over time he now knew him to be a genius above all the many brilliant artists and artisans who had stepped forward in the rebirth of European art since the worst effects of the Black Death had begun to abate.

Verrocchio believed that Leonardo would – among his many achievements in many fields - change visual art forever.

In 1472, the two men had begun a collaboration on a painting for the Church of St. Salvi in Santa Verdiana. The painting, "The Baptism of Christ," Began as Verrocchio's commission, and he was to have been the primary painter. But Verrocchio was happier sculpting than painting. Among the other apprentices who would work on the painting, the very young Leonardo – still barely more than a teenager - was to have helped with the background and one of two figures in the mid-range; …an angel holding the cloak of Jesus.

On seeing Leonardo's work on the painting, however, Verrocchio was overwhelmed to discover that Leonardo had absorbed all of the stylistic brilliance and artistic breakthroughs of the visual artist Giotto in his paintings of a hundred years before; Leonardo was now extending Giotto's hard-won knowledge into new directions in Verrocchio's very own workshop. With this early example of Leonardo's genius, Verrocchio was delighted beyond all measure.

Even so, Verrocchio's working relationship with Leonardo was usually very troubled. Leonardo was a relatively tall man; very imposing, head-strong, and rude to subordinates even when young. Many of the apprentices were angry at his abrupt manner of pointing out defects in their work. Fortunately for Verrocchio, they were able to mind their manners because they also recognized the precious gift of Leonardo's close scrutiny. A few carried grudges and sought revenge in ways that Verrocchio found to be very disruptive and difficult to resolve.

There was no question that Leonardo had many great insights and ideas. Following the introduction of Guttenberg's printing press in Germany several years before, Leonardo's use of now much cheaper and more commonly available paper for sketching, rather than parchment, was proving to be a godsend. This was true not only for Leonardo, but for the entire arts community in Florence. On the other hand, in considering Leonardo's proposal for using linseed oil as a medium rather than egg tempura for the pigments in fresco, Verrocchio believed the change to a new medium will cause many problems. Even so, on the chance that Leonardo might have something with the use of oil, Verrocchio was willing to stop arguing with him about his use of it.

Leonardo was not intimidated by the artistic achievements of the workshop's master. Nor was he impressed by the high regard for his master's work by the most powerful family in northern Italy, the Medici. Nor was Leonardo intimidated by deadlines, or the need to wrap up a project because the client was threatening to withhold payment for the work. Leonardo did his own work at his own pace.

Even so, despite his preference for working alone, he was recently named as a Master Artist by the Guild of Saint Luke. The certificate of the Guild in hand, and with the support of his father, Leonardo had recently set up his own workshop. He was beginning to understand the larger problems of business ownership, and the need for apprentices and bookkeepers who shared his artistic vision without complaining about the hard and boring detail work often assigned to them.

Verrocchio was able to talk briefly with Leonardo a few times over the next several weeks. The two men were associated on several projects, and there was a constant need for talk. When Verrocchio first mentioned the young girl and the interests of their patron, Lorenzo, Leonardo was dismissive; ...as though he did not want to be bothered.

Some days later, Leonardo himself brought up the subject in a discussion of client business. "The young woman you mentioned, Andrea. Have you decided to have her come here so that we can have a look at her?"

"I am surprised to hear this from you. I thought you would reject the idea. The girl – her name is Orabella – has impressed our patron with her skills both in the arts and in supervising the work of several of his household staff, but I have not met her, and I have not seen any of her sketches. Nor have I heard further from Lorenzo's messenger, Adolfo. I was prepared to send a message to Lorenzo that we could not use her. But I wanted to hear your thoughts before I said anything at all. Should I ask Adolfo to send her over with some of her artwork?" Verrocchio asked.

"I doubt if fine clothes, art work, and an introduction by a personal representative of the Medici will be a good introduction for her to our group here; or ours to her. Better, I think, if we introduce her as the daughter of an anonymous friend's family who would like to give her a chance to earn some money, and, maybe, learn something about drawing, painting, and sculpting. Introduce her as a kitchen helper and tell her she has a month to show her worth.

"I am amazed and pleased, Leonardo, that you have given this so much careful thought. I applaud your reasoning. In the spirit of getting her directly involved with our work here, I will propose to Adolfo that we will send someone to escort her here by way of the market place so they can pick up some food for the household. That way we can get some idea of how she will work with our cooking staff, as well as some of the marketplace thieves who supply us.

"An important test. Who would you propose as an escort?" Leonardo asked.

"Vincenzo, the pigment grinder, has more experience in the market than anybody here. He is probably the best judge of the talents of a female buyer. He will probably want to pick up some pigments as well. A purchase of a quantity of Lapis Lazuli will certainly tell us something about her innate sense of the value of precious minerals.

"That will give us a lot of information to make our judgements on her suitability for the workshop, and for the probable hostility toward her from the apprentices."

"Would you consider Amadeo as escort?" Leonardo asked.

"The Plaster maker? What value do you see in that?", Verrocchio responded.

"Listen my very good friend," Leonardo paused briefly, then said "Since starting my own workshop, I have discovered that many of the commissions I receive include requests for a variety of small frescos. These are used as gifts, and for various household decorations.

"I know very little about frescos and have little interest in making them.

If your girl doesn't work well in your workshop, she may work very well in mine because she won't have to put up with the contending factions among the apprentices. If she is as fast a learner as you have suggested, then, within the month, she should have enough knowledge of plaster making, paint mixing, and fresco design to produce some of these minor artworks. In that role, she could be enormously valuable to me.

Verrocchio thought a moment, then spoke. "As always, Leonardo, you have amazed me with the breadth and depth of your vision. Plus, I am greatly relieved that you see yourself with a major stake in this girl's ultimate success. I will be more than happy to lay this out with Adolfo. I am sure he will be discrete and will carry it to Lorenzo Medici with enthusiasm."

"Good. By the way, have Amadeo ask the girl – did you say her name is Orabella – pick up some sweets for the apprentices on her way over here. Maybe that will make them more receptive to her charms." "Consider it done, Leonardo," Verrocchio said.

====

Amadeo's job in Verrocchio's workshop is the management of everything to do with the making of plaster for frescos, and pigments for paints. He was a little surprised to be assigned to escort a new kitchen helper to the workshop, even though she was to be escorted from the Medici household, perhaps the most prestigious address in all of Florence. He might have been insulted that he had to also escort her through the food markets on a shopping trip, except that he was intensely loyal to Verrocchio, and would do whatever he asked with all the talent and ability that Amadeo could bring to it.

"You look surprised, my friend," Andrea Verrocchio said to his long-time apprentice, Amadeo. "I'm told this is a very pretty and intelligent young woman, though still a bit coarse in her manners. Even so, when you present yourself at the door of the Medici, you will have a chance to actually see the inside of the home of the most powerful family in all of northern Italy, as well as the most important and generous of all the patrons who support our work."

"Yes, Andrea. But I am a little overwhelmed with it. You know that I am only one of the helpers here. I really have no training and no experience with this kind of work. Wouldn't Leonardo be a much better agent for you? I know he is a favorite of the Medici family."

"First of all, Amadeo, you are not just 'one-of-the-workers-here.' I place great value on your experience, and your work with the plasters and paints. Without your work too many of our apprentices will be sitting around wondering what to do with all their pretty sketches and designs because they will have no materials to express themselves with. In the situation with this young woman, you have more talent and interest to bring to this particular task than does Leonardo himself." Amadeo waited, wide-eyed.

Verrocchio went on. "As you know, Leonardo has learned a lot about fresco during the years he has been with us here, but he has little interest in working with it. To be truthful, neither do I, but I do recognize that just about everyone in Florence with a piece of blank wall wants to have one of our plaster fresco panels. They will want the panel framed in wood with brightly colored images of The Virgin holding the baby Jesus and surrounded by cherubim and seraphim.

The family wants a fresco that will last forever, that will hang on that wall in the family house, covering up that blank space. I also know that we can create those panels, in quantity, in our workshop, under your supervision, but you will need more help."

Verrocchio continued. "Now that Leonardo is trying to build business for his own studio, he and I both agree that this girl may be able to learn enough fast enough, to help us all produce more of these small frescos before our competitors recognize the value available in this market."

"But if you and Leonardo are prepared to invest so much in a girl that you have never met, wouldn't you both want to escort the girl yourselves so as to get to know her better?", Amadeo asked. "She will have to be tested in some way, and quickly I assume, before her value can be judged at all."

Verrocchio paused before answering. 'You would think so, wouldn't you. The problem is that we want this girl to start out as a kitchen helper and maid, and we will have to avoid giving her any impression of special favor by either myself, or Leonardo. If the girl is to be of any value to either of us, she will have to rise on her own. I am assured by Lorenzo through his messenger Adolfo, that she is fully capable of that if she has proper, and very subtle, encouragement and guidance.

"I would expect her to make herself available to you within a couple of months of starting her work here. And, when she does, you must be ready to assign your most disagreeable, dirty, and dangerous tasks. And you must be prepared to guide her through them so that she learns useful things without putting the health and safety of all of the rest of us, and the workshop itself, in great danger."

"So, my friend, perhaps the most important part of your escort duty will be to impart the most interesting aspects of purchasing materials like limestone for plaster, and other minerals for grinding into appropriate colors for fresco. If she seems worthy, let her work closely with you in the purchase of a quantity of Lapis Lazuli. I have a commission coming up that will require more than we presently have on hand."

Verrocchio continued, "Perhaps you could let her actually select and purchase some mineral of lesser value. Your report of that, particularly on the quality, amount, and price of the actual purchase, will give us more understanding of Orabella's real talents than anything Leonardo or I can glean by talking with her."

"Didn't you also want me to take her to do some shopping in the market for food supplies for the house?", Amadeo asked.

"Yes, I did," Verrocchio responded. "Why don't you do something like telling her that we have to buy some meat and vegetables for the evening meal, and that it will have to feed all of the household staff and apprentices; ...perhaps 25 people in all. Tell her that she is to decide what to purchase, and how to get the best price. If she does it to your satisfaction, give her the money to make the purchases. I will discuss all our plans with our cook, Giulietta. The cook will also be contributing her share of difficulties that will further test the mettle of this young girl.

"After all, she will have to survive the kitchen, before she will even get a glimpse of the workshop."

Amadeo spoke again. "One last thing. You know that I have a bunch of randy male apprentices working in my shop. A young and pretty girl sounds like trouble to me. Do we have any idea how she might respond to one of their typical rude, pushy comments toward newcomers? Or do we have any idea what might happen if one of those boys decides he wants to check out her sexuality and gets a little pushy."

"I regret that I do not. Leonardo and I have discussed that possibility, and I have already expressed my concerns about those possibilities with Lorenzo's messenger. Adolfo believes that she can take care of herself, and he says he is speaking with the full authority of Lorenzo. Just be alert. At the very first indication of trouble along that line, be prepared to bring me into it, but make sure to give her plenty of time to work it out on her own.

"You may have to find work to do that keeps you close to the studio where she works for the first several days of her time there."

"I am relying on your judgement in all of this, Amadeo. You are a good man, and I believe you will do the right thing. Your work will make all of us proud. I now have other business to attend to. As soon as I have made arrangements with Adolfo for the date and time for you to appear at their door. I will alert you. If you have further questions on this please let me know. Goodnight for now."

"Good night Andrea."

Verrocchio departs the room. Amadeo sits for a moment contemplating all he has learned this day.

===

Verrocchio understood the surprised look on Amadeo's face as he described the escort task to him. He hoped that his request that he also take Orabella through the market for the raw material used for pigments needed to make paints, would remove any concerns Amadeo might have about his status in the marketplace. Verrocchio knew that he needed to replenish his store of raw lapis lazuli for a new commission to be done mostly by the very talented apprentice named: Vincenzo. Lapis Lazuli is the raw material for making the very precious ultramarine pigments used in commissions for their wealthiest patrons.

Lapis is expensive; ounce for ounce about the same as gold in the Florentine jewelers' market. In 1475 Florence a day's food ration of bread, wine, and meat totaling 2,000 grams for a laborer would cost about 24 denari, or 2 soldi. One ounce of gold or the best quality Lapis would cost almost 1,000 soldi; …the equivalent of nearly 500 days of food for one man or one woman.

Amadeo took quill to paper to estimate the total amount of money it would cost for the girl's shopping trip and mineral purchase training: 25 staff for dinner at 2 soldi each = 50 soldi; ½ oz of lapis lazuli at 500 soldi = 500 soldi; plus 50 soldi for miscellaneous tips and minor purchases; a total of 600 soldi. With an exchange rate of 1 florin to 120 soldi, Amadao will ask Verrocchio for 5 Florins to cover costs of the escort.

With his plan and budget in place, Amadeo only had to wait for Verrocchio's message with the requested time of Amadeo's arrival at the Medici home. He shivered in anticipation and began to worry about his ability to do the job his master had asked of him.

====

Several days later, Amadeo appeared at the main double-doors of the Medici house. He straightened his shoulders, approached the great wooden doors and rapped sharply. As the doors opened he felt his knees grow weak. He worried, briefly, that he might faint. He breathed a sigh of relief as Adolfo, Lorenzo's personal messenger, appeared and said "Welcome, Amadeo," and waved him inside.

"Wait here, Amadeo, while I see if Orabella is ready. She does want to make a good impression on you."

Amadeo is a shy man. The idea that a pretty young girl would want to impress him is a little unsettling. The idea that she would accompany him through the very crowded and very public Florentine marketplace did not give him peace. Nevertheless when Adolfo brought the girl from a room, Amadeo resolved to stiffen up, and carry out his duties.

"Amadeo. May I present Orabella, a respected member of our household staff. We have very much enjoyed our time with her, and almost regret that she is leaving us. We take heart that we know the Verrocchio household is a great opportunity for her, and we wish her well."

Orabella was dressed in neat but plain clothes, suitable for a housemaid about to go outside the house to do household business. But she carried herself with authority. She approached Amadeo boldly with her hand outstretched, as if to shake hands with him. Amadeo had never heard of such a thing from a young woman. His words of greeting stumbled out of his mouth, but her bold approach so overwhelmed him that he put out his hand to hers and shook it.

"I am pleased to make your acquaintance, Orabella. Are you ready to apply your skills to the House of Verrocchio?" Amadeo was pleased with the recovery of his dignity. Perhaps she didn't notice his initial discomfiture.
"And I am pleased to make yours. I am ready. May I call you Amadeo?"
"Yes. Of course." Amadeo said.

Adolfo spoke up, obviously pleased with this initial encounter: "We will send a cart over to the workshop with Orabella's belongings later today. Is there anything more I can do for either of you?"

Amadeo bowed, as did Orabella. Both offered a parting expression of warmth and gratitude, as virtually any Florentine would offer in response to an expression of favor by a senior member of the House of Medici.

Just as he was about to turn away, Amadeo thought he noticed the wife of Lorenzo, Clarice Orsini, looking on from a high balcony. He took her interest as a good omen.

====

Amadeo decided to start his tour with Orabella in one of the dockside shops along the Arno where larger rocks from the distant quarries were broken down into smaller rocks suitable for specialized purposes, usually in the arts or other decorations. These smaller rocks could be further broken down and crushed by Amadeo, or one of his apprentices, for use as pigments when mixed in some kind of medium for use in colored paints. Pigments made from minerals were very important in frescoes because minerals – unlike any plant or other biological material - would not degrade the bond between the dried plaster and the paint color. The degradation resulting from the use of plant and animal materials in paint will ruin the fresco over time.

Later that day, Amadeo and Orabella were in a shop of one of the dealers in meats for that night's dinner for the staff of the workshop. The shop was in an upper story along the Ponte Vecchio looking northeast along the River Arno. The shop's owner, and principal butcher, Anselmo, greeted Amadeo warmly because he represented Verrocchio, a prized client of his butchery.

"Greetings to you Anselmo. This is Orabella. She will be working with the kitchen staff in the Verrocchio household. I'm sure you will see more of her over the next several weeks. Please know that she will represent Andrea as well as I myself."

"I am very happy to meet you, Orabella. Bella, Bella, Orabella. Amadeo. How do you rate the pleasure of the company of such a beautiful woman?" Before Amadeo could answer, Orabella spoke: "I'm very happy to meet you, Anselmo. I can only hope that your meats are of the highest of quality standards, as are your kind words, while preserving the most modest, yet reasonable of prices."

Anselmo was dumb struck by her words. In his world, no woman, and certainly no girl like this one, no matter how pretty, talked this way on first meeting a gentleman shop-owner.

Anselmo looked in Amadeo's direction and winked. "Your young friend has a tongue, Anselmo. Should I always expect this kind of remark whenever she calls on my good feelings toward your master when she buys meat?"

Orabella immediately apologized. "I am very sorry sir. But I am new in this neighborhood. In my former neighborhood, I have had to take a strident attitude when dealing with new shopkeepers, in order to establish the respect necessary to do my Master's business. I can see that you are a different kind of person, a gentleman I can deal with. I apologize to you as well, Amadeo, I did not mean to cause problems with one of the household's primary, and most respected suppliers."

Amadeo could not help smiling, though he was most discrete to avoid letting Anselmo see his enjoyment. "I'm sure we will all survive this encounter, Orabella. Let us find out if Anselmo can forgive your impertinence when you place an order for 2,500 grams of meat to be delivered this afternoon."

"Of course, Amadeo. I will instruct my staff to wrap one of my best cuts of meat for delivery this afternoon. Be assured that there will be no hard feelings between myself and your young woman. Besides only a fool would feel anything but joy at the opportunity to work with such a pretty and forthright customer."

Before anything more could be said Orabella spoke again directly to Anselmo. She was smiling broadly and showing all of the charm she was capable of thanks to the training in manners, dress, and presentation that she had received in the Medici household: "May I examine the meat before you begin cutting. I would like to see how you go about your business."

Anselmo rolled his eyes and looked toward the ceiling, then to Amadeo, then back to Orabella. "Of course you may. Please come with me."

Finally, their business completed, Amadeo and Orabella bid Anselmo a warm goodbye until next time. They went on their way to their next stop on the docks west of the Ponte Vecchio along the Arno. Amadeo needed to order quantities of various rocks needed for the colored pigments used in frescos.

Amadeo believed that the most important part of his job escorting Orabella, in the eyes of his Master, Verrocchio, will be the knowledge he is able to impart to Orabella in his negotiations with the masons and rock smiths who cut and price the relatively small quantities of rock he needed for pigments.

The rock smiths were big men who were used to dealing with other big men in very physical ways. Though it looked to Amadeo that Orabella would do fine in her dealings with them, it was clear in Amadeo's mind that the butcher shop owner, Anselmo, had been a push-over compared with one of these tough men.

One of them approached. He looked up and down Orabella's body. Though her long skirt hid her feet and ankles, her face and her upper body and waist under her coat displayed an obviously attractive young woman. Though she blushed a little, she did not turn away from his too lingering gaze. Finally, he said to Amadeo: "Yeah?"

Amadeo did not recognize the man, though he had done business with this rock yard several times over the years. They had always had the stock he needed, and the price was always acceptable. "I am with the House of Verrocchio. This is my apprentice, Orabella. We need to purchase some small quantities of several varieties of rock to crush up for colored pigments that we will use in the artworks we do on commission.

The smith looked toward the ground and began to scratch his cheek. When he looked up he seemed confused. "What kind of rock do you need," the smith asked.

"We can start with 20 kilograms of limestone, then I need to see your stocks of ochers, siennas, umbers, terra verte, and any compounds of iron, or manganese and iron that you have available. I need some greens, so If you have malachite, I'd like to see that."

The smith said "My name is Alfredo. Come this way. It may be muddy. The girl can wait here if she wants."

"She is my apprentice. She will come with us," Amadeo said to the back of the already walking away Alfredo.

"I'm pleased to meet you, Alfredo," Orabella said to Alfredo's back. "I hope we will soon be doing business with you."

Alfredo looked back over his shoulder but said nothing. The raw limestone was rough, but of sufficient quality to satisfy their purposes. Amadeo and Orabella both picked through the other minerals and compared them for purity and quality. Once they had what they needed they asked that their purchases be delivered to the house of Verrocchio sometime within the coming week.

Alfredo quoted a price for the entire order, including delivery. Amadeo and Orabella conferred for a moment then Orabella offered 15% less than Alfredo's quote. "I'll pay you 25% of the total now, and 75% on delivery," Orabella said.

"Alfredo's face got red. He bent toward Orabella and said through clenched teeth: "I have offered you a good price. Do you think I am trying to cheat you?"

"No, Alfredo. Not at all. It is just that the rock samples you have shown us are of a lesser quality than I am used to working with. If you have samples of better quality, I will be happy to look at them, and will be happy to reconsider my offered price if I believe they actually are of better quality." Alfredo looked toward Amadeo, but Amadeo had gone some distance away to look at some different kinds of rock. "Wait here," Alfredo said. He walked toward a small building that might have been the rock yard owner's office. Shortly, he returned with another fellow who was not as big as Alfredo, but with a more commanding presence. "This is Matteo. He owns the yard and will be happy to discuss better quality samples and prices with you."

"It is my pleasure to meet you, Orabella, I think that is your name, yes?" Matteo asked.

"It is my pleasure as well, Matteo. Do you have some better samples I can look at?" Orabella asks.

"Yes, I do, but you will need to come into my office to see them."

Orabella hesitated at the invitation, but only for a moment. Amadeo was some distance away, but Matteo did not appear to have any other motives toward her than selling some of his products to her. She followed as Matteo walked ahead of her toward his office a short distance away. Once there he went into a back room and brought out a tray of small samples of more carefully selected stones for her to look at.

Satisfied with their quality compared to the samples shown her by Alfredo, Orabella had no problem arriving at a price for each, and for a time for delivery. Matteo also expressed no concern with her proposal for an initial payment today, and a final payment to complete the transaction on delivery. At the end of it, Orabella offered her hand to Matteo to seal the deal.

Amadeo had given some part of the purse for today's purchases to Orabella so she could complete the purchase of the minerals to be used for pigments. She made the initial payment to Matteo, then went looking for Amadeo.

One more purchase remained, that of the very precious Lapis Lazuli, only available as an import by camel caravan from the Hindu Kush mountains in Afghanistan, a several month-long journey overland to Florence from the east. Lapis Lazuli is ground into ultramarine pigment to be used in painting the ceremonial robes of kings, emperors, and the Virgin Mary. Orabella had asked, but Matteo did not have the security necessary to protect a stock of Lapis. He referred her to a jeweler near the Medici bank who might be able to provide her with the precious material.

Amadeo had been walking among some of the piles of materials while Orabella conducted the business of the Verrocchio household. When she caught up to him he had paused to feed some bread to a raven that had perched nearby. The raven put a foot on the piece of bread and tore off a more edible chunk with his beak, then crouched and looked up from his eating to note the arrival of Orabella. She noted a white feather on the upper right wing, and asked Amadeo how long the raven had been with him.

"For several minutes now. The raven is a very intelligent animal, and this one seems to want to talk after each morsel of bread." After Orabella told him of her purchases, and her experiences with Matteo, he asked, "How did you learn how to judge the quality and price of these lesser stones?"

"In the House of Medici, there are always people available who can examine and evaluate materials brought to the household for sale. Because they believed me to be nothing more than a curious girl, - who might also have the favor of Lorenzo - those people would not hesitate to answer my innocent questions about their testing and evaluation methods."

Orabella continued, "One of the Medici people knowledgeable about the price of Lapis today says that the 12th part of a libbra will have a current value of about 8 Florins, but you must check the daily price on the Florins to be used for the transaction before agreeing on price. As I am sure you know, the real problem in determining value in any sample of raw precious minerals are the inclusions that take away value, and the purity of what remains after the inclusions have been removed."

"We have dealt with the same dealer in Lapis and other precious minerals for many years," Amadeo said. "I am sure he is trust-worthy, but if you are able to test a sample of Lapis for purity in a way that our dealer will agree with, we might be able to get a better price. That would please the Master."

"I have never done such a test," Orabella said. "We will have to buy the Lapis today at the dealer's price. Once we are back at the workshop I will try to find out more about how to do the testing to see if we are getting good value." ----

The two shoppers completed their shopping list and were able to return to the House of Verrocchio before nightfall. Following the instructions of the Master and of Leonardo, Amadeo introduced Orabella to the main cook in the kitchen, Giulietta, and to the Master of the household staff, then took his leave. Giulietta gave Orabella a clean but worn dress and apron and showed her to her small bed in a corner of the large pantry. Once changed into the work clothes she had been given, Orabella was to come back to the kitchen and help with cutting the vegetables, and with the cooking for the evening meal.

Orabella was not surprised at her treatment as a common kitchen helper. Giulietta was very picky and critical in finding fault with Orabella's work. Though Adolfo had not told her what she would be doing on her first days and weeks, he did say that she should do every assigned task with the same selfless and energetic enthusiasm as she had when she was first brought into the House of the Medici almost two years ago.

"If you can do that with good cheer," Adolfo had said earlier that day as he walked with her to the entryway where Amadeo waited, "I have no doubt that you will rise quickly to more important work in the House of Verrocchio."

Orabella approached her assigned kitchen tasks in the way Adolfo had hoped. Even on that first day, she could see some changes in the work flow that might please the cook, but she resolved to wait until the meal had been served, the diners satisfied, and the clean-up completed. If there was any kind of discussion of what went good, and what bad during this dinner, that might be a time for suggestions.

Or maybe she would wait for a few days before saying anything. She was sure the household staff would test her in some way – possibly many ways - and she would need to survive all their tests in order to gain their respect before making any suggestions for changes in their workways. She allowed herself to become encouraged and hopeful about this testing period when she noticed that Giulietta, in the midst of a highly critical comment about the poor result of a particular task assigned to Orabella could not quite suppress a smile.

Orabella worked hard with the staff. They may have imposed tests on her, but the tests did not seem to take away from her general cheerfulness. After several weeks both Giulietta and the staff supervisor looked forward to working with her and seeking her suggestions for how they might improve their work and get more of their work done in less time. As good reports of her efforts began to be laid before Verrocchio, he began to think it might be time to bring her into the workshop to see if she could work the same kind of magic with the apprentices. He will discuss his plan with Amadeo, first, and then he will take a walk over to Leonardo's studio for further discussion.

A few days later Verrocchio came to Amadeo's studio. "I am told, Amadeo, that our new girl is working out very well with the household staff. Have you heard anything to the contrary?"

"No, Master. She is everything Adolfo has told us. If you want to bring her into the workshop, I have no objection. When the time is right for her reassignment, I think it will be best to have her start by cleaning up the fresco studio and organizing the paints and materials in there. That will bring her into contact with the apprentices, and we will see if there is going to be any problem there. I will give her plenty of room to work out any problems with the apprentices in whatever way she thinks best. I am very confident that she will do well."

"When it is time," Amadeo continued, "I will choose a well-qualified apprentice to work with me on the new fresco commission. Orabella will assist us. If she works as well with the apprentices as she has done with the household staff, she will have my highest recommendation., and my full support in whatever she may choose to do while she is here."

"Thank you Amadeo. You are a good man. I will discuss all of this with Leonardo. I am sure he will agree to our plan." With a wave toward Amadeo, Verrocchio turned to go back to his private studio elsewhere in the workshop.

====

Amadeo looked around his studio to contemplate, once again, the process of making a fresco. The English word for fresco is "fresh" to indicate the nature of the wet plaster that is the base layer. The base layer is to be followed by two additional base layers of plaster. The final layer – called "intaco" - is the painting layer. The application of the artists' design and colors must be completed before the painting layer is dry; usually within several hours. Together, over a period of many days, all these things together will become the final painting. The techniques used to produce a fresco are ancient, simple, and well-understood. However, because of the complex interplay of factors in the quality of materials selected; in the processing of these raw materials; in the sequences involved in the application of the several layers of plaster and paints; and in the drying times between applications; the mastery of fresco techniques can be very daunting.

It begins, basically, with rendering limestone into a calcium carbonate by cooking it in an outdoor kiln at a very high temperature (1000 - 1100 C) for a few minutes to remove Co2. The carbonate – now called quicklime - is alkaline and very corrosive. It will burn skin if touched. Quicklime is then rendered into a hydrate by adding a bit of water. The result is called "slaked" lime. When dried, slaked lime takes its final form as a white powder.

When a suitable wooden frame has been constructed for the fresco, or an area of wall has been selected and prepared to receive the first base layer of plaster, the slaked lime is mixed first with small amounts of water and then with a quantity of a sand such as volcanic ash in the proportion of 8 parts sand to 5 parts slaked lime.

It is the proper mixing and application of the sand and slaked lime where Amadeo's experience and expertise came to bear. He is confident in his ability and in his ability to pass his skills along to Orabella.

=====

"I am very pleased to hear that the young girl is working out so well with your household staff," Leonardo said. "If she is able to deal with Amadeo's randy apprentices, and actually starts working on a client fresco, I will want to see her work in progress. Make sure that Amadeo gives her significant work in the preparation and mixing of color pigments, and in the preparation of a design that can be shown to the client. I will be interested in seeing the quality of her work in those most difficult of fresco problems, and I will criticize any work she does accordingly."

"I am very confident, Leonardo, that she will pass any test of her artistry that you may want to give her, though she has not claimed to be an artist, nor has Adolfo given me any reason to believe that she may have talent in the area of visual design. We may find that she does have that kind of talent, or we may find that she does not. In either case the real question for our Orabella is whether she can use the resources of the studio to produce frescos of sufficient quality and quantity to satisfy the customers who are willing to buy them from either of our studios.

"Are we agreed on that?" Andrea asked.

"Yes, Andrea. We are agreed."

A week later, Amadeo came into the kitchen, begged Giulietta's forgiveness for taking Orabella away from her chores, and asked Orabella to walk with him to the fresco studio. The cook and the cook's helpers all wished her well, gave her hugs, and asked her to come back to the kitchen whenever she felt the need to visit, or to eat some meat, bread, and special sauce.

Before leaving with Amadeo, Orabella went to her curtained off space in the pantry to change into the rough cloth pants, smock, and hair cover that Amadeo had given her to wear in the studio. For now she would return to this space when her work in the studio was done for the day. Eventually, if things worked out, she hoped she would have a more private space closer to the studio.

Amadeo was clearly embarrassed as he talked to Orabella about the general harassment and sexual harassment she was no doubt going to suffer, initially, from one or more of the five apprentices who worked with frescos. "They are quite full of themselves," Amadeo said. "and their occasionally rude and offensive behavior quite beyond any criticism in their eyes. But I want you to criticize them to their faces when they do not do their work. The only authority you will have will be your femininity, and your sexuality and your talent. That makes you different from them and may gain you some tolerance from them that you would not otherwise have if you were another male. Do you understand me?"

"Yes, Amadeo. I know I am to be challenged if I am to have any real opportunity to continue working in your studio. I believe I can stand up for myself and bring honor to this house. Thank you for giving me this opportunity."

====

The apprentice challenge to Orabella's sexuality came within a few weeks and was as quickly resolved. For her part, Orabella took extra care to button the studio clothes and wear them loosely, and to bind her breasts so as not to give any ideas to the apprentices. She was well aware that the making of frescos is a very physical process with many reasons to lift heavy objects and push and pull them into place. Working with the kiln also generated so much heat that the clothes still on the body could become almost transparent with sweat. No matter how well she conducted herself there would be sexual temptations all through the day in the fresco studio.

She had dealt with such temptations in the times before she was taken from the streets by the Medici, and she had learned how to deal with most of them, though not all. Amadeo had told her to report any threats to her by any of the apprentices, but Orabella knew that such threats could never be reported unless her life was actually in danger. Any such complaint that got back to the apprentices in any of the studios would mean the end of any possibility that she could eventually gain their trust and confidence in her work with them. Even their tolerance for her continued presence in the studio would be seriously at risk.

Orabella did have one confidant in the person of the First Cook, Giulietta. Giulietta had been so impressed with Orabella's conduct during her time in the House of Verrocchio, that they both became close friends. Giulietta had been in the house for almost 5 years. She had an intimate knowledge of the internal politics among the staff and could see that Orabella would need help in dealing with what Giulietta knew to be some hostile feelings between some of them.

Antonio Sforza had been brought into the House of Verrocchio as a favor to Lorenzo de Medici.

Lorenzo was, no doubt, working out some kind of arrangement to solve one of the many problems that occasionally arose with the House of Sforza in Milan since the Sforzas had taken over The House of Visconti some 20 years before.

Lorenzo's machinations always involved lands separating the Medici lands from those now owned by Sforza. Lorenzo's machinations would often entangle Verrocchio. As always, Amadeo was willing to help his Master by taking in young men like Antonio as an apprentice even though he could not immediately see any useful way of putting him to work in the fresco studio. Coming from the Sforzas, Antonio had a certain attitude of superiority over other apprentices, and a lack of deference toward Amadeo. Rather than try to learn some aspect of making frescos within the mentorship of Amadeo, Antonio preferred to occupy himself with learning about, and engaging in household politics among the staff.

It turned out that one of the apprentices in the fresco studio, Poldi, a helper in making plaster who had more looks and muscle than intelligence, took a bet offered by Antonio. Poldi said that he could persuade Orabella that she should have sex with him, and she would agree. Antonio bet that she would not do it, knowing that Antonio's challenge to his manhood would give Poldi added strength in actually convincing her to do it with him. Antonio promised to assist in the enterprise by finding a reason to persuade the other apprentices to leave the studio on some kind of special task needed by the Master.

When the designated day arrived, Poldi told Orabella that she needed to help him fire the kiln in order to render a batch of limestone, into quicklime for a new commission. Orabella was always eager to help with these kinds of very specific tasks so that she could gain as much experience in as many aspects of the workshop's works as possible.

"I have to go back into the studio for some tools. Then I'll go to get some wood to get the fire going. Use the wheelbarrow to start bringing the limestone over there and put it next to the kiln," Poldi said.

Poldi went into the studio, caught Antonio's eye, and indicated that the time is now. In a few moments, the studio was empty. Poldi returned to the wood pile, took his shirt off, picked up an ax, and began splitting the wood into pieces that would burn with a fierce flame. "Orabella!" He called. "Come help me with splitting the wood when you have finished moving the limestone."

Orabella did not hear what he actually said, but she had heard him say something. She went to the back of the woodshed where she knew Poldi was working. Since he often took his shirt off when working with the kiln, she was not at first concerned about it when she saw his bare chest.

"This wood is still green. I need some help separating the ax from it, and some help getting the split wood out of my way. Can you do that?" He asked.

"Yes of course, Poldi. Stand out of the way, and I'll clear the wood behind you."

As she walked behind him, he turned and put his hand on her shoulder. "You know I have had my eye on you, Orabella, since you came to work in the studio. I think under those loose clothes you are probably…"

That was as far as Poldi got in making the proposal that he was sure would be successful and would ultimately win the bet with Antonio. In the space of an eye blink, Orabella had turned toward Poldi, and had kicked him, hard, in the testicles with the solid toe of her shoe. Poldi screamed and fell to the ground rolling, crying, and screaming. He pulled his knees to his chest while holding onto his painful private parts. Orabella knelt beside him. She told him to get up and stop blubbering.

Poldi was barely able to sit up. His crying turned into a whimpering moan. He continued to hold his testicles, and he could not sit comfortably because he was in a lot of pain. Tears were streaming down his face.

"You should consider yourself lucky. I could easily have put out one of your eyes. You made a mistake, but I will forgive you for the insult if you will promise me this: do not ever talk about this to anyone, and especially do not talk to Antonio. Do not try to claim that you won the bet. If he asks you, tell him he will need to talk to me directly to find out what happened. I will dispose of Antonio in my own way. If I ever hear that you have discussed this misbehavior of yours with anybody, the swelling in your balls might never go away, and you might never again find work in the city of Florence.

There is one more thing. If any of the other apprentices ever approach me with the same idea that you had a few moments ago, then you must walk over that person and hit him, hard, right in the face, with no warning. Then walk away and go back to your work. Can you do that for me?" Poldi – still crouched over, moaning and sniffling - nodded yes.

"I mean you no harm if you behave yourself, but if there are more of these challenges waiting for me anywhere in the workshop, I will need your help. Do we have a deal? If so, put out your hand." Poldi did so. Orabella shook his outstretched, but shaky hand. "OK. It is time to get up when you are ready. We have to finish preparing the limestone."

Amadeo had come into the studio to find out why all the studio apprentices were away when there was so much work to be done getting ready for the new client. As he looked for someone, Orabella and Poldi walked into the studio from the Kiln. By the look on Poldi's red, teary face, and Orabella's calm demeanor, he knew exactly what had happened. His first thought was that he was glad the apprentices had got their sexual challenge to Orabella out of their system, and nobody had been badly hurt. Poldi would recover. "Hopefully," he muttered, soto voce, "we can now focus on the work… at least until the next problem comes up."

====

Once the sand has been thoroughly mixed with the quicklime and water, and the backing surface has been prepared with a rough texture, it will have to dry for 1 or 2 days, or until the backing layer is tacky to the touch. The first layer of plaster is then laid on. That layer will also have to dry for 2 days, or until it is felt to be "ready." Then a second layer of plaster is laid on and allowed to dry. A third layer may also need to be applied, but this will be up to the judgement of the fresco's visual artist. He or she must decide when to lay on the design.

=====

Leonardo and Verrocchio were very pleased at the reports they had received about the confrontation between Orabella and one of the apprentices in the fresco studio. When thinking about the expression on Poldi's face when he first walked into the studio with Orabella to face Amadeo, both broke into laughter.

"I believe Orabella has earned the right to supervise the creation of a fresco under the watchful eye of a paying client," Leonardo said.

"I agree, "Verrocchio said. "She will still need some training in design. Amadeo has not yet had an opportunity to see if she can work with the client to come up with an acceptable design. At the point where she has developed a design for the client's inspection, I think it will be important that you come in to inspect her work before it goes back to the client for final approval, and interim payment. Agreed?"

Leonardo nodded, yes. He went on to say: "Stay in touch with Amadeo as she works her way through all the steps in preparing the plaster for the application of design. I want to know if she has any particular problems developing any part of the fresco, from crushing the minerals to make pigments, to laying on the plaster layers, and allowing them to dry to the correct feel.

======

A few days later Amadeo gave Orabella authority to supervise the apprentices in the creation of the new fresco. When he left the room, she asked them to gather round to discuss the course of the work. She called first on Poldi to ask about the status of the current batch of calcium carbonate; rendered shortly after the incident behind the woodshed. Antonio Sforza had been banished from the workshop, and Poldi had learned to be respectful and deferential around Orabella. She, in turn, had become courteous toward him, and in her supervision of him.

"The quicklime will be ready for mixing whenever you are ready, Miss," Poldi said. "If we need more, Fons will help me fire the kiln. We have plenty of limestone."

"Good. Thank you, Poldi. You are a very good helper to me."

Orabella was already beginning to feel guilty about the way she had dominated Poldi. She wondered if she had misunderstood his intent. She knew that Antonio Sforza, the lazy trouble-maker, had put him up to it, but she was beginning to doubt that Poldi would have gone through with it. Orabella directed her next comment to Vincenzo, the mixer of pigments: "Are there any problems grinding any of the rocks we need for the pigments, Vincenzo?"

"No, Miss. As soon as Fedele and I have the cartoon with the design and colors, we will be better able to judge the suitability and quantities of the materials we have on hand," Vincenzo replied. Then he asked: "Pardon me, Miss, but who will be doing the design?"

"I am told by Amadeo that Leonardo, though he does not have a particular interest in working with fresco, has taken an interest in this one. When the time is right, I am told, he will work with us on a design that our customer will find acceptable. In the meantime, I will need to work with Elia on the creation of the cartoons that express the artist's design. Elia will work directly with Leonardo as he always has, but Amadeo has asked that I learn something of the creation of them in anticipation of future commissions.

Vincenzo spoke up again: "When he last spoke to us, Leonardo expressed some interest in exploring the use of linseed oil as a medium for the pigments rather than egg whites. He believes the oil – because it takes much longer to dry, weeks instead of minutes – will give him much more flexibility in composition.

"The problem is that visual artists have been using egg tempura on parchment for at least a thousand years. We are very experienced with it, and I am a little concerned about Leonardo's wish that we break from these ancient traditions."

Orabella responded. "Thank you for expressing your concern, Vincenzo. This is an important question. Have you done any work with linseed oil as a color medium?"

"Yes, Miss. I have mixed two batches of pigment using the oil because Leonardo requested it. If he is now asking to use oil for this commission, I guess he was satisfied with the earlier results. "

"Has either Amadeo or our Master expressed an opinion on it?" Orabella asked.

"No Miss. My understanding from both those gentlemen is that they want to give Leonardo as much freedom as possible in experimenting with new techniques, materials, and methods. That is, of course, unless the client specifically forbids experimentation beyond traditional practice. In this case, to my best knowledge, the client has not mandated adherence to tradition."

"Do you know of any particular problem with the use of egg tempura that suggests that we should fully embrace the use of linseed oil in the pigments?"

"Some have suggested that frescos that have relied on any compounds using organic matter like eggs or plant materials lose their color and tend to grow mold in that part of the fresco where they are used. I have not observed these problems myself, but they do give me pause in advocating their use here. Linseed oil is also made up of plant material, so I'm not sure what we will gain by using it in fresco.

Orabella was thoughtful for a few moments. "I don't want to make a decision on this, but I do want to keep the discussion going at least until we can hear more from Amadeo. I will ask Amadeo to confer further with the Master, and with Leonardo. Hopefully, I will have a better answer within a few days. For now, I would like to get to work.

"Vitale. When can you have the frame done for our new fresco?"

Vitale responded: "The frame can be ready later today, Miss. We have followed the client's preferences for a frame about 3 feet by 4 feet."

"Good. Poldi. Let's meet early tomorrow morning to look at Vitale's work. If the frame is ready I want to lay the first binder coat of plaster in the frame so it can begin to dry."

=====

The key to success with fresco is the quality of the plaster, the timing of its' application, and the time allowed for drying each separate layer. The size of Orabella's fresco made it relatively easy to work with the complex application of layers and drying times. If the painter had all the pigments prepared, and then worked steadily with a good plan of execution, he or she could finish painting the 12 square foot area in about the same amount of time required for the plaster to dry. If the plaster is allowed to dry, then any subsequent applications of pigment will no longer be fresco, because the wet pigment would no longer bind chemically with the dry plaster.

The last of the three binding layers would have a roughened surface to provide the best binding surface for the "intaco," or the actual painting layer. Because of the necessary drying times, creation of the binding layers would require a little over 2 weeks.

While waiting for the binding layers to dry, Orabella spent her time working with Elia on creating the life size cartoons from the finished work of various artists available in the studio. Cartoons are line drawings and sketches that represent the artists' designs for the fresco in outline form. Cartoons are scaled up, or down from the artist design, to match the scale of the fresco itself. Holes are punched along all the lines in the cartoon.

Once the painting layer has been applied and allowed to dry for about 20 minutes, the cartoon is positioned according to the plan for the completed fresco, and then temporarily taped over the wet plaster. The lines in the cartoon are then "pounced" with a small sack of crushed charcoal. These "pounced" lines are the boundaries to be observed with each application of pigment.

Application of pigments must be completed within the drying time of the "intaco" layer; …about 12 hours depending on temperature and moisture in the air. Once the pigment is fused chemically with the plaster by drying it can never be changed. This chemical binding is the secret of the longevity and freshness of a fresco over thousands of years, but it can be very frustrating for the artist who feels the need to do a "touch up."

=====

Leonardo had come into the studio when Amadeo told him that Orabella was ready for his scrutiny. He came to the studio in good spirits and left in even better spirits. He was so pleased with Orabella's work on the cartoons prepared from the studio's stock of paintings, that he offered to help her do the actual design work for the fresco. When he heard of the interest of Leonardo de Vinci in the work, and his offer to help the manager, Orabella, with design of the fresco, and with any other problems she might have to solve, he offered to double the amount of the commission.

This made both Verrocchio and Leonardo so pleased with Orabella's work, that they assured Amadeo that her time with the artists workshops of Verrocchio and Leonardo, in the City of Florence, would never end. On hearing of these comments, by way of Amadeo, Orabella knew she had to make amends to Poldi. She told Amadeo of her regret at the treatment of Poldi following the incident at the kiln. She resolved to apologize to Poldi. She would work with him to make sure he could restore his good reputation with the other apprentices.

PANEL 7 - THE NEW WORLD

Eirian, a bright young woman trained, surreptitiously, in advanced mathematics at the University of Edinburgh, Scotland begins work with James Clerk Maxwell at his rural estate in Glenlair. They will test and document the partial differential equations that Maxwell has developed to define electromagnetism. She studies the writings of Ada Lovelace and attends an off-beat lecture by Charles Dodgson.

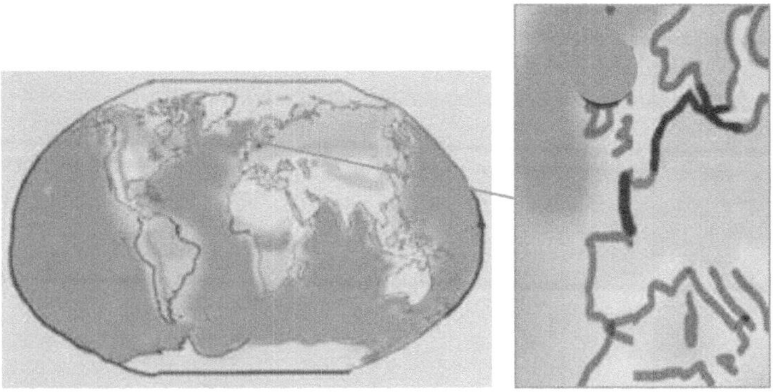

EDINBURGH, SCOTLAND. WINTER 1865 CE

PROTAGONIST

Eirian Ross - A sturdy young woman born and raised on a farm in rural Scotland.

MINOR CHARACTERS ASSOCIATED WITH EIRIAN.

Bethan Sutherland - Eirian's closest friend from school and church in Corsock.

Sean McCabe - Freshman student in Dodgson's math class at Oxford. Assigned by Dodgson to escort Eirian and Bethan around the campus during their visit to attend Dodgson's lecture.

Ada Lovelace – (b. 1815 CE – d. 1852 CE) An early inspiration to Eirian. She was a designer of flying machines and programming languages.

James Clerk Maxwell – (b. 1831 CE – d. 1879 CE) is considered by many today to be the third most important figure in the development of physics and mathematics after Newton and Einstein. He is a religious man who becomes an Elder of the Church of Scotland.

Michael Faraday – (b. 1791 CE – 1867 CE) discovered electro-magnetic induction. Faraday was poorly educated but his discoveries of electromagnetic induction became the basis for the extensive work of James Clerk Maxwell to further develop the science and technology needed for electric motors, electric utilities and all modern electronics.

MINOR CHARACTERS ASSOCIATED WITH MAXWELL.

Thomas Sutton – (b. 1819 CE – d. 1875 CE) Collaborator with Maxwell on creation of the first color photo in 1861.

Rene Descartes – (b. 1596 CE– d. 1650 CE) Philosopher of the scientific method; creator of the systems for graphing algebraic formulas onto three orthogonal axes called Cartesian Coordinates. Each point in such a system can be located by the distance of the point from the 0 point on each of the three axes: (x, y, z).

Charles Babbage – (b. 1791 CE– d. 1871 CE) Inventor and builder of a prototype for the first steam-powered computer. A later design for an analytic and programmable version used a programming methodology developed by Ada Lovelace. Babbage held the Lucasian Chair – once held by Sir Isaac Newton - in Mathematics at Cambridge from 1828 to 1839.

Charles Dodgson – (b. 1832 CE– d. 1898 CE) Also known as Lewis Carroll, author of "Alice's Adventures in Wonderland", published in 1865.

MINOR CHARACTERS ASSOCIATED WITH DODGSON.

Henry Liddell – (b. 1811 CE– d. 1898 CE) Dean of Dodgson's college at Oxford, Christs' Church; father of Alice Liddell, Dodgson's model for the character of Alice in "Alice's Adventures in Wonderland."

Alice Liddell – (b. -1852 CE - d. 1934 CE) Daughter of Henry Liddell. Presumed model for Alice in "Alice in Wonderland."

It can be argued that the coordinate graphing system with x, y, and z axes orthogonal each to the other, otherwise known as the Cartesian Coordinates, is the most fundamental of all the tools of applied mathematics. 17th century Philosopher Rene Descartes developed his system of three coordinates as a part of his life's work to bring formal methods to the process of scientific discovery. Toward the end of that century, the Cambridge alchemist, mathematician, lens-maker and astronomer, Isaac Newton, uses Descartes' coordinates to develop a formal calculus to determine, with mathematical precision, the orbits of the planets around the sun.

By the late 17th century, Guttenberg's printing press has already enabled these and many other new, radical ideas to be widely distributed in the form of paper handouts and cheap books. The authority of the Catholic Church is weakened by the effects of plague, and by the widespread challenges to Catholic doctrine and practice by many churchmen associated with the thinking of Martin Luther in Wittenberg, Germany.

In the 18th century French Philosophes Voltaire and Rousseau speak to the need for new ideas of social justice, for greater awareness of the natural world, and for scientific proof based on experiment. Newton's calculus, with the notational refinements by Leibnitz in Germany, enables a membrane to form between the traditional workings of faith and the evolving workings of formal science and social awareness.

Thus begins the Enlightenment.

It is late fall, 1865. Four years earlier, James Clerk Maxwell, then 30 years old had demonstrated the world's first color photograph. Now, he has just resigned his Chair of Natural Philosophy at King's College, London to return to the family farm at Glenlair, Scotland. While there, he will perfect and publish the ground-breaking mathematics that form the basis for his studies of light and electromagnetism. Maxwell's studies will become the foundation for the future development of the quantum mechanics needed for radar, computers, and cell phones in the 20th century, and needed for the ways to unleash, and harness, the enormous binding power within the structure of atoms.

Edinburgh is cold this morning, and Eirian Ross shivers as she waits for the small coal stove in Dr. Maxwell's office to heat up. Dr. Maxwell had sent her up from his private office on his family estate at Glenlair, a day's train ride away, to do some research in the University library. She reads a copy of

Edinburgh's "Scotsman" newspaper, scrounged from a desk in the office, while she waits for some tea water to heat up enough to drink. The office has been provided to Maxwell as a courtesy by the University of Edinburgh in recognition of Maxwell's graduation from the school, and his likely future contributions to Physics following on to his brilliant work as an undergraduate at Edinburgh, then at Cambridge for advanced degrees, and, most recently at King's College, London.

Eirian was born and raised in Corsock in southwest Scotland. Her parents were teachers and farmers, and Eirian's house was full of books. Their conversations at dinner were full of grand ideas, travel, and detailed mathematics. Maxwell's family at Glenlair lived nearby. She was well acquainted with Maxwell – who is often called by others, though never by Eirian - by his nickname, "Dafty." 13 years her senior, Maxwell is a personal and family friend, a teacher, and as a mentor and physicist, who, her parents told her, is probably destined for greatness.

Eirian smiles as she remembers Maxwell's attention to her. From her earliest memories, he has always treated her like a good friend even when young; and now, very much like a professional colleague. She remembers all the puzzles and games he would bring to her attention when she was younger. As she stumbled over the challenges presented by these entertainments, Maxwell would offer suggestions on how to work through the most challenging problems. For the very most difficult problems of all, they would work together. Eirian remember these moments, especially, with great fondness.

Around the age of 12 her parents decided Eirian had progressed much further in her learning, than they could keep up with by home schooling in Corsock. They asked for Maxwell's help in getting her enrolled in the Edinburgh Academy, the school he himself had attended.

Eirian was not happy to remember that she experienced her first rude awakening to life's realities when Maxwell came to their home in Corsock to tell them that enrollment in the Academy would not be possible. "The Academy is an all-boys school," he told them "and very protective of that tradition."

"Surely he could see my disappointment because I could not hide it, any more than I could hide my tears as he spoke the words," Eirian whispered to herself. "My parents were downcast, tears were flowing down my face, and all was quiet in the room for some moments."

Eventually, then, Maxwell spoke. "Let me think on this. You all, as a family, should continue to think in terms of sending Eirian to Edinburgh. I am now a senior Professor at the University there, and I have complete access to all the teaching resources of the University. I have developed good relationships with several members of the teaching staffs, and I believe I can encourage them to act as her tutors from time to time.

"For myself, I will continue to oversee Eirian's course work, and will agree, here and now, to read and grade all of her papers. The important thing, however, is her moral upbringing. She is a pretty girl, and, with her obvious intelligence, she will grow into a very attractive woman. I am not able to give her guidance in dealing with the many temptations that will come her way as she matures.

"I think it is very important that you both accompany her and continue with the most important work of parenting and encouragement in her studies until she feels she has gained enough confidence in herself, and awareness of the world to be on her own. You are both excellent teachers, and I can help you find appropriate employment, introductions and housing around the University community."

Maxwell concluded by saying: "My family will take care of your properties at Corsock while you are away. You will always have your home to come back to."

====

Eirian recalled that day in Corsock as being one full of the extremes of emotion, even at her age of twelve years. She and her parents walked with Maxwell back to his carriage. The sun had broken through the usual clouds and drizzly rain typical of the area through the year. The rolling green hills suddenly exploded in the distance, filled with sunlight.

She felt a little guilty as her mind wandered from the greatness of Maxwell, her boss and mentor, to the thought about the long walk from his offices to a stand where the horse-drawn cabs usually wait. Eirian had not slept well in the rooming-house near the campus the night before because the difficulties of Maxwell's assignment roiled her brain. Eirian rubbed her eyes at the thought of another night of a few hours of restless sleep… if, indeed, she would enjoy any sleep at all.

Eirian's brow furrowed as she picked up the notes on the mathematics of the statistical problem that Dr. Maxwell had asked that she bring with her so she could do the necessary research in the University Library. In particular, he has asked her to carefully examine his recent paper titled: "A Dynamical Theory of the Electromagnetic Field." She knew she would have to build her skills in partial differential equations and statistics, and she would have to gain some level of understanding about his formulations of the so-called "operators"; 'div' (the divergence of a flow); 'grad' (the gradient of a flow); and 'curl' (the amount of twist in a flow).

These concepts were the result of his deep thinking about the "shapes" of forces at work when concepts of electricity are joined with concepts of magnetism in ways that Michael Faraday had investigated and published in his 1859 book "The Forces of Matter". Eirian knew that these concepts would be at the heart of any of Maxwell's future writings on electro magnetism.

Eirian tried, initially without much success, to envision in her mind what these terms might mean in describing shapes occurring in space as though there were no physical connection with the three-dimensional world she could see with her eyes, touch with her hands, and smell with her nose. Maxwell had suggested that by envisioning these shapes turning freely in space she would more easily see how to apply the mathematical formulations he had developed.

Maxwell believed that Newton had developed his calculus in this way by envisioning the shapes of the forces that connected the planets. Eirian found this idea of experimentation occurring in the laboratory of the mind rather than the laboratory of flasks, beakers, sinks, and gas jets profoundly difficult, but she resolved to continue until she got it right because that is what Maxwell wanted. She knew that that was also what Maxwell needed if he was to communicate his complex and not-yet-popular ideas to a less scientific public who might sit in judgement on his funding requests, or even to the more scientifically literate members among his professional colleagues who might be persuaded to support his efforts if he could persuade them.

Eirian also knew that one of the reasons Maxwell had brought her into his laboratory was his appreciation for her deep curiosity – almost a match for Maxwell's own – and her probing intelligence and facility with math, but also for her willingness to get her hands dirty in the lab. Maxwell enjoyed reminding her of the time he had come to visit her family, and she, barely six years old, was down in the mud trying to pull a screaming piglet out from under a pile of wood where his tiny hoof had gotten stuck, so she could re-attach the small animal to her mother pig's nipple.

With these images in her mind, Eirian could not help but look at her fairly coarse hands, toughened and strengthened along with the rest of her body during the years she spent growing up among the animals and the endless hard physical chores on the farm, and the summers she spent there when not in school.

Maxwell's concerns about education and schooling were limitless. Whenever in Glenlair, he worked tirelessly to establish a school and church in his small community. While at Marischal College in Aberdeen, he had established Working Men's Colleges and had conducted regular lectures to those who would become engineering aides, builders, mechanics, draftsmen, and administrators. They would have the responsibility for assembling and maintaining the structures housing the electrical and steam machines that would power the Industrial Age.

Maxwell's concerns about the need for technical training across the entire spectrum of the Industrial Age was not limited to the factory floor and the machine shop. He also held regular weekly lectures and hands-on instruction for University students who would then be able to build the lab equipment and do the lab work and calculations in support of the work of scientists like Maxwell.

Maxwell's strength as a scientist – as yet unrecognized by many of his professional peers - lay in his refusal to fully accept ideas that could not be tested. Maxwell's scientific thinking on electricity and magnetism, for example, was far ahead of most of his scientific colleagues in that they still held firm to the concept of a mechanistic, clockwork universe as shown by Sir Isaac Newton almost 200 years before. Maxwell's testing of the ideas of Michael Faraday, and the frequent electrical shocks he got when the testing went astray, had begun to give him profound doubts about this dogmatic, though traditional view of how the universe actually works.

For Maxwell, testing also meant that he had to not only devise the tests, but, more often than not, he had to design and build the test equipment in his workshop in order to do testing on new, unorthodox ideas that had never been tested before.

Eirian looked around the little office at the pieces of testing gear that Maxwell had brought with him when he left King's College, London. Of all of these pieces, Eirian found herself most interested in Maxwell's color top. This device was a step along the way to a separation and definition of the spectrum of colors contained within sunlight, a step that would lead eventually to the world's first color photograph. Maxwell presented the strategy for producing such a photograph to a meeting of The Royal Society in 1861. Eirian thought aloud to herself. "It is not that Maxwell is a hard task master. He is a fundamentally decent man who is totally engaged in the work he has set out for himself. From an early age through the present day, with the firm and dedicated guidance of both his parents, and the full support of his wife, Katherine, he has been a dedicated student of the Bible and he is an Evangelical Presbyterian. But unlike other churchmen, he has a passion for exploration of the fundamental forces of the universe. His mind works at the level of "vibrations" between molecules. "By his passion and enthusiasm, he has fully engaged me in the work as well," Eirian continued.

"He has asked me – when I have some free time - to look over the draft of his paper 'On Reciprocal Figures, Frames and Diagrams of Forces.'" Eirian knows that this paper is a step beneath the level of his work on electricity and magnetism. Though the paper applies to Maxwell's other interest in architecture, and the details of building construction, Eirian could not help but think that there is a larger message here. "This is a method for the formulation – the "forming" if you will - of general solutions to problems involving Newton's 'Action at a Distance,' which is the heart, soul, and primary motivation for all of Maxwell's investigations into electro-magnetism. Eirian thought to herself. "I am very privileged to have been given this assignment."

Eirian resolves to never allow herself the luxury of a mistake when she develops her own work on projects Dr. Maxwell has requested of her.

Eirian is reminded that her working relationship with Maxwell this time is somewhat complicated by the fact that Maxwell is living and working mostly at his family's estate in Glenlair, about 80 miles southwest of Edinburgh. There, he is only partly concerned with advanced mathematics. Mostly, he is engaged with his family in a plan of architectural work and development on the main buildings on the estate in line with the vision of his father, John, who had died in 1856.

Though born and raised in Corsock, a village about 9 miles north of Glenlair, Eirian now works often on the campus in Edinburgh so she can become thoroughly familiar with the Library and those of its' extensive reference materials that are directly relevant to Maxwell's work. Occasionally, Eirian brings some reference materials and mathematical formulas to Glenlair because she needs Maxwell's close attention in the solution of some problem.

Sometimes Maxwell returns to Edinburgh to meet with banks, carpenters, and stone masons involved with the expansion work at Glenlair. Sometimes he goes to sit through a peer review of his current scientific work, and to discuss any changes in the direction of the development work that are indicated by his most recent findings. Occasionally, Eirian accompanies him.

Usually, Eirian's editing work on Maxwell's drafts is done in Glenlair where Maxwell is close at hand to consult on test results, wording, and presentation. Sometimes he returns to Glenlair without prior warning. He knocks on the door of her work room at the family estate to announce that he is back, and asks "...are you ready to go over your work with me?"

Eirian knows that Dr. Maxwell receives a lot of skeptical looks when he introduces her to his professional colleagues who might not have met his wife, Katherine. After all, there were no women enrolled in a degree program at the University of Edinburgh because women were not allowed. They wonder why he did not choose one of the many very bright, very talented young men in physics or math from the rich academic communities in Edinburgh, or, indeed, from any of the many institutions in Scotland. Also, because she knows that she is attractive to many of those same young men because of her status with Maxwell, she sees the rather "knowing" looks she gets from older women and men outside the science community when Maxwell introduces her to them. Eirian long ago gave up being concerned by these provocative, but also, often, rude and insulting social cues.

=====

In her life course, Eirian imagined that she only had to set a priority on her study of each of her interests, and a future date when the study would begin. She could not believe her good fortune in being hired by Dr. James Clerk Maxwell upon his return from his recent position as Chair of Natural Philosophy and Astronomy at Kings College, London. Even better, he has invited her to help him with his work in the most important scientific inquiry of the 19th century: the space-time formulations describing the movements of atomic particles, and the propagation of light and energy that are at the heart of his life's work on light and electromagnetic radiation. She has come to believe herself to be a true scientist; not because of any academic training – the University of Edinburgh does not accept women as undergraduates, and the whole idea of women studying science is generally unthinkable anyway in these days – but because her curiosity – aided by the constant support of her parents and the support, mentoring and tutoring of Maxwell - about all the mysteries of the universe is unquenchable; her readings in the sciences are voracious, diverse, and deep, particularly in Mathematics; ...the language of science.

Being blocked from access to higher education by the most prestigious University in Scotland caused Eirian to begin talking out loud to herself in this moment. "The University of Edinburgh does not knowingly accept a woman as an undergraduate, but they had unknowingly accepted a woman in 1911," she whispered to herself.

Earlier, in July of this very year, Edinburgh's daily paper, the "Edinburgh Scotsman," published a story about the death of James Barry, an 1812 graduate of the University of Edinburgh Medical School. Following academic examination by the Royal College of Surgeons of England in July 1813, Barry went on to a long and successful career as a military surgeon in the British army.

A broad smile broke across Eirian's face as she recalled the article naming James Barry as the deceased person, then referring to Barry by her real name - the name she was born with - Margaret Ann Bulkley. Bulkley / Barry was a woman who had also borne a child before creating an entirely new male personae for the sole purpose of gaining entry to a degree program at the University.

On reading the article in July, Eirian had resolved to fight in whatever way she could for a woman's right to earn a degree from the University of Edinburgh. From that July day forward, the Margaret Ann Bulkley story would be the cornerstone for all Eirian's future efforts to see a woman enrolled and matriculating as a four-year student at the University of Edinburgh. That woman would enroll as a woman, not as a bright and intelligent girl forced to masquerade as a boy in order to get an education.

Through her infrequent readings of the daily "Scotsman" newspaper on these trips to Edinburgh, Eirian slowly became aware of a group of women who had been gathering formally and informally to talk about women's issues and problems. They hoped to provide a point of contact for others who supported the ideas of women's suffrage, and the more specific idea of gaining full access to the University of Edinburgh for women.

A spin-off group advocated for entry into the curriculum leading to a degree in medicine from the University of Edinburgh. They were becoming known as "The Edinburgh Seven," and Eirian resolved, in that moment in July, to use her connection to the prestigious Maxwell to support the efforts of these women.

Though Eirian's academic talent seemed to lie in the realm of abstractions of science and mathematics, her up-bringing on the family farm at Corsock put her in constant contact with all the animals of the farm and all the problems of farm animals – animal feeding, animal wastes, diseases, still births, abandoned calves, lambs, and chicks; these were the everyday occurrences on the farm. Even as a very young child she had on a few occasions found herself with her arm buried in the uterus of a pregnant animal to re-arrange a potential newborn into a safe position for birthing.

She had, long ago, lost track of the number of headless, bloody chickens she had been required to dip into hot water and de-feather so that her mother could cut them up for dinner. The smell of wet chicken feathers was still vivid in her mind, though, she was happy to say, it had not kept her from eating the succulent birds.

In this rare and extended session of personal wool-gathering, reading about James Barry brought Eirian's mind to the whole issue of slavery. She thought, too, of the recently concluded war between the states in the American Colonies. Great Britain, through its Parliament, had finally, and peacefully rendered the institution of slavery illegal in 1833.

The first major crack in the British slavery edifice occurred with a judicial finding in 1783 that slavery within England itself was no longer permitted by law. However, an anti-slavery law that could reach all the way across the far-flung British Empire around the globe could only occur after a period of intense legal and political battle over the issue following the 1783 finding.

Ultimately, in English civil society, the growing awareness of the destructive and terribly wasteful uses of human life under slavery in the distant English colonies could no longer be tolerated by the English people, and the 1833 act of Parliament making slavery illegal had been the result; …all done without bloodshed, at least within the home islands, Eirian noted to herself.

Eirian regretted that It had taken another 33 years for the issue of slavery to be resolved in America by the defeat by force of arms of the slave-holding southern states, and then only after a terrible and bloody civil war. At the thought of the more than 600,000 war dead in the Colonies, Eirian felt a shiver go through her body.

She also thought it galling for all the people of the British Empire that the cotton-growing plantations and mills in the southern American states had been the primary suppliers of cotton for the clothing mills in England, and in London. In particular, English manufacturing and banking interests in London in their passion to acquire enough cotton for their mills were no less morally corrupt than slave-holding southern plantation owners.

For the independence-minded Scots the English material support to the Confederacy in the Americas was a major source of irritation in all their dealings with the powerful commercial and political interests in London. Now that London's commercial bet on the Confederacy has been lost in the defeat by Lincoln and his army of the Northern States, the irritation across Great Britain grew deep enough to cause the English liberal party's most influential member, Gladstone - a slave-holder who had engineered the support for the Confederacy - to lose his influence to the politically conservative party of Disraeli. Disraeli – a close friend of Queen Victoria - had maintained his neutrality on the issue of aid to the slavery interests in the American Confederacy, and was able, with Derby as Prime Minister, to pick up the reigns of government with the collapse of Gladstone's political majority.

Eirian thought briefly about the fate of US President Lincoln; assassinated two weeks after the defeat of the southern army. Shot in the back of the head where he had sat with his wife Mary in Ford's theater by a pro-Confederacy actor named Boothe; himself later captured and killed.

Rather than get more upset, Eirian pushed her thoughts of slavery aside, and opened a publication containing a series of Christmas lectures given in December 1859 by Michael Faraday titled: "The Forces of Matter." The lectures are dense with ideas and calculations about force fields around magnets, around pieces of shellac, and about the polarity of electrical currents, etc. Maxwell considers Faraday's work to be fundamental to his own work. Now that she had warmed her body with the heat of burning coal and hot tea, and her mind with her passions about the most urgent social issues constraining humanity and humankind, she knew it was time to begin work on the task Maxwell had assigned her.

Eirian Ross is a bright, hard-working farm girl with a life-long curiosity about the things around her. She has a deep urge to tinker, to explore, and to fix things. She was born, and has been raised, in Corsock, a small farming village in rural Scotland a few miles north of the estate of James Clerk Maxwell and his family at Glenlair. Her parents, both teachers, had a small piece of farm land, a barn, and a house flooded with books. Eirian was encouraged from a very early age to take full advantage of them. With the help and encouragement of her parents, once she began reading she never stopped.

Her parents lived modestly, and with considerable frugality so they could save up for travel. Whenever possible, the Ross family took the overnight train from Dumfries station to Edinburgh, stayed with friends who worked at the University there, and often accompanied Eirian on her forays into the great libraries. They traveled extensively around Scotland, and, once every few years, to London, and to the great Universities at Cambridge and Oxford. Eirian was always with them. She absorbed all the sights, sounds, and smells of all the distant places visited. Upon return to Corsock, she could recite the names of all the places, and all the extraordinary sites they had seen and visited.

As she grew out of childhood, she began to read more deeply into those subjects that truly interested her, like the mathematics she had found useful in solving puzzles and riddles on cold winter days. On the warm sunny days, she enjoyed laying in the grass so she could more easily envision the three dimensional physics of the wing motions that supported birds in flight. As with many children growing up in rural areas with lots of nearby woods, the colors, the songs and chatter, and the flights of birds became fascinating.

One prominent woman writer known to Eirian, Ada Lovelace, though many years her senior, had as intense an interest in birds as did Eirian at the same age. Ada lived far away in Cambridge, England. She kept careful notes of her observations of birds in a booklet later published under the title: "Flyology."

Any book written by a young girl that had anything to do with the science of those days held little interest for most scientists; However, Eirian's neighbor and fellow churchman who lived a few miles south on the road from Corsock, James Clerk Maxwell, had no such inhibition. He was well-aware of the many interests and extraordinary talent of his young neighbor to the north. He had been her mentor in all things academic since she was barely more than a toddler. Now that she is an adult he is happy to make sure that she has access to whatever papers, scientific or otherwise, that she took a fancy to.

Maxell became aware of the "Flyology" paper because of his scientific work with colleagues of Charles Babbage at Cambridge, while he was at King's College, London from 1860 to 1865. Between the years 1828 and 1839, Babbage had held the Lucasian Chair in Mathematics there, as had Sir Isaac Newton 150 years before. Babbage had also developed a design for a steam-driven "Difference Engine" while still in his late 20s and early 30s. The Royal Society and the English government had acknowledged the huge time savings over the otherwise error-prone and time-consuming calculations by hand of extensive tables of values used in all manner of commercial, scientific, and military endeavors. They had committed and spent many thousands of English pounds trying to help Babbage and his machinist, Clement, make the gears and cogs that would bring his design for an automated calculation system to life.

If the difference engine could be made to work it would speed up the calculation of tabular values like logarithms and tables of bearings and distances needed for ship navigation and land surveying. This was the product The Royal Society had paid for, and the very least they had hoped for. Though eventually built in prototype, the machine never did work as intended mostly because the tolerances in the manufacturing processes this early in the 19th century were not nearly well-developed enough. Ada

Lovelace worked closely with Babbage at Cambridge in the design of an Analytical Engine which would overcome the problems experienced with the difference engine. She had translated a paper written by a young Italian Engineer named Luigi Menabrea following a lecture at the University of Turin by Babbage on the Analytical Engine.

Lovelace's Notes on her translation were longer than the translation of Luigi's article. One of her notes proposed the use of a programming model for the Analytical Engine based on the Jacquard Loom. The Jacquard Loom used a pattern making system for making flowers, leaves, and other complex organic patterns in English cloth. Lovelace developed a new language using the Loom as a model. Ada's model would give scientists and others programming tools they would need to create sequential processes on Babbage's Analytical Engine. Once these sequential processes were loaded on the engine it would process each step, in sequence, to solve complex problems unique to their particular line of work, or endeavor.

Unfortunately for Babbage, the failure of his attempt to build a working Difference Engine, and the refusal of the British government to finance any more attempts, meant that Babbage could continue his life of lively social rounds among the glitterati of London and Cambridge, but could no longer find the funds to turn his new design for an Analytical Engine into a workable product.

Even worse, perhaps, was the death of Ada Lovelace from cancer in 1852 at the age of 37. Ada's programming language for the Analytical Engine could not be truly tested without a machine to test it on. Scientists like Maxwell continued to wonder about the implications for society had Babbage and Lovelace been successful in developing and building a programming system designed to take advantage of all the capability of Babbage's fully functioning Analytical Engine.

=====

Whenever in London, Maxwell took every opportunity to meet with Babbage and the aging Faraday to discuss the future of science in this new industrial age based on coal, the manufacture of steel, and the technology of the steam engine, and the telegraph.

Eirian's vivid imagination often returned to a visit to the Oxford University campus in July of 1862. She and her childhood friend from Corsock, Bethan, had been invited by Maxwell to travel with him to that Ancient and prestigious University to attend a summer lecture by Charles Dodgson on the Mathematics of Euclid. Rumor had it, according to Maxwell, that Dodgson intended, someday, to put the Axioms of Euclid into a Shakespearean theatrical format; each axiom would start a new scene.

Maxwell thought Dodgson's idea was preposterous, but also thought the lecture would give an interesting insight into the weird, wonderful, and magical mind of Charles Dodgson.

Dodgson was a sometime mathematician, many times a story-teller, always brilliant and totally creative in bringing ideas from far afield into his ambit as senior lecturer in Mathematics at Oxford.

Maxwell had great confidence in the math skills, and the intelligence and drive of his young protégé, Eirian, and he wanted to give her every opportunity to expand her scope of knowledge into the realms of fantasy and reality that the tight logic of Mathematics could not reach. Maxwell hoped she would continue to help him work on the documentation and testing of the differential equations he had been developing to describe the deep forces at work in electro-magnetism while he was a Professor at Kings. Dodgson's skills as a mathematician were not stellar, and his ability to present and lecture was erratic at best. But he was a story-teller who fashioned wonderful and fantastic characters, figures, and situations to bring dimensions, color, smells, and life to his narratives. Maxwell was a churchman and an Episcopal elder, but he was also a scientist who had created and demonstrated the first color photograph. He was now in deep scientific exploration and mathematical description of the mysteries of electricity and magnetism; magic that wanted to break traditional physics, and Mathematics free of Newton's mechanistic universe.

Maxwell knew there was a connection between heaven and earth, and he had dedicated his life to finding it. Despite the Scottish conservatism in his point of view, Maxwell sometimes let his mind wander along the pathways suggested by the imaginative creatures and fantastic events in Dodgson's story-telling. Maxwell had other business with colleagues Babbage and Faraday in Cambridge and London on this trip, so he excused himself after the Dodgson lecture to catch a train for the 65 mile train ride to Cambridge.

Despite his stuttering, and his fairly clumsy presentation, Eirian and Bethan decided to see if they could get some time face-to-face with Dodgson. They knew little about him except for the interesting stories that Maxwell had told them. Though all about mathematics, Dodgson could not seem to restrain himself when he searched for an animal reference to use in his commentary. He sprinkled hookah smoking, talking blue caterpillars, disappearing Cheshire cats, waist-coated white rabbits, and officious dodo birds with abandon throughout his lectures.

After the lecture, Eirian and Bethan approached Dodgson with a couple of carefully thought out questions, but before they could pose them, he immediately invited them both to join him and some of his friends at the rowing ponds on the Oxford campus. Dodgson's friends included H.J. Liddell, Dean of Christ Church, Dodgson's College; Liddell's wife, Loreena; and his daughter, Alice.

As an afterthought, Dodgson invited one of his bright young freshmen in Mathematics, Sean McCabe - who had attended the lecture - to act as a guide and interpreter for the two young women who had come down from Scotland with Maxwell at the invitation of Dodgson.

Sean was a nice looking young man, but he seemed very shy. When he introduced himself to Eirian and Bethan, he stuttered. "H-, h-, hello. M-, m-, my name is Sean. I'm, m, m happy to meet you," he said.

Eirian had been working to take the hard edges off the deep Scottish brogue she had inherited from her parents and her farming community in rural Scotland, but it was still strong enough that it would not fool a member of the English upper class like Sean. Because Sean stuttered, however, Eirian hoped he would be sympathetic to her difficulties with speech and dialect on the grounds of this most traditional of English institutions.

"Hello, Sean. It is very nice to meet you. My name is Eirian, and this is my friend Bethan. We are from Corsock in southern Scotland, near where Dr. Maxwell lives. Will you walk with us?" "Y, y, y, yes, " Sean replied.

Later as Sean stepped away to talk with someone, Bethan turned to Eirian and whispered: "Sean is a very nice looking young man. Do you think he will someday get over his stutter and his shyness?"

"Yes. I think so. Perhaps he stutters because his mentor Dodgson stutters," Eirian replied.

As the day on the rowing pond evolved into an afternoon of tea and conversation, Eirian took note of Dodgson's fondness for the young Alice. She was not then more than 10 years old, but very pretty, charming, and quite mature for her age. She was somewhat infatuated with Dodgson, as he was with her.

"They seem quite infatuated with each other," Eirian said to Bethan as they looked upon Dodgson and Alice. Dodgson had found a rabbit hole among the roots of a large tree near the stream where rowers passed by in their punts. Alice leaned over to look into the hole, and it looked for an instant that she might fall in. Dodgson caught her by her waist and pulled her back out. She looked at him and began to giggle.

Dodgson stepped a few paces away and sat on a nearby bench. He looked toward Alice, then motioned her to come over. He pulled a sheaf of bound papers out of his satchel. She sat down beside him as he began to read.

Eirian spoke to Bethan: "Before we decided to come down here, James told me that – in addition to the business at Cambridge - the other reason he wanted to come down to Oxford was that Dodgson had just completed a manuscript for a book of fantasy, and James thought I might enjoy meeting Dodgson. He told me that I can invite you along for company if I want, and he will take care of the costs.

"He told me to think of the trip as a bonus payment for the editing and testing work I have been doing for him on the equations. I am glad you decided to come with me, Bethan"

"Look at the father. He seems a little uncomfortable with their behavior, " Bethan replied.

Eirian looked in the direction of Alice's parents for a moment. "But not so much that he wants to try to break it up." She said. "I wonder what's in that manuscript."

"Maybe more strange caricatures of mother Goose Rhymes, or Grimm's, or Hans Christian Anderson Fairy Tales…like in his lectures," Bethan said. Then, in an aside, she concluded, "I'm very happy you invited me to come with you, Eirian. Thank you." They held hands for a moment, then Sean returned, and suggested that they join the others for refreshments.

Eirian posed a question to Sean: "Do you see the manuscript Mr. Dodgson is reading to Alice?"

"Yes. Why d, d, d, do you ask?"

"Do you know what it is about?," asked Eirian.

"Y, y, y, yes. Yes, I do. Mr. Dodgson is writing a book of f, f, f, fantasy that features Alice. He is concerned about her reaction to it, so he reads from it, to her, at every opportunity."

=====

The recently published writings of Michael Faraday were also important to Eirian, though not because she was personally interested so much, but her mentor, Dr. Maxwell, insisted that she become familiar with Faraday's work. In Faraday's lab work in the 1830s – Results had been published early in the previous year, 1865 - he had developed ideas and demonstrations around the concept of fields of force such as those that curve around to connect the positive and negative poles of a bar magnet. These lines of force can be persuaded to show themselves by laying a piece of paper over the magnet, sprinkling some iron filings on the paper, then lifting slightly and lightly shuffling the paper while moving it around above the magnet.

During the shuffling, the filings begin to fall into the curved lines that represent current flow – the force fields - between the two poles of the bar magnet. These curved lines represent "action at a distance." In other words, the action of one object transmitted to another object without an obvious connecting medium. Understanding these lines of force and the methods of transmitting force from one object to another across what appeared to be empty space was very important to all the work in electricity and magnetism now being done and described by Maxwell himself.

Faraday used his findings to develop ideas of a magnetic field around a conductor carrying a direct current. Faraday established the basis for the concept of the electromagnetic field in physics. These studies in turn had evolved into the electro-magnetic spectrum that Maxwell had investigated and further developed with methods in advanced calculus while a department head in Mathematics at King's College, London.

Further development of these concepts had now become Maxwell's life's work. Testing, documentation, and publication of these findings would consume much of Eirian's time over the next few years. As planned, Maxwell would, alongside his wife Katherine, oversee Eirian's work. In this period dedicated to the testing of his experimental work on electro-magnetism, and the formal presentation of it to the academic community, he, himself would concentrate much of his personal attention on the further development of the Glenlair property according to the vision and wishes of his Father, John.

====

Maxwell's Demon.

Maxwell had asked Eirian to focus some part of her studies of mathematics on statistical methods. When asked by Eirian why these methods would be important, Maxwell replied that he was trying to account for the actions of individual molecules in a gas.

"Obviously, such molecules, in their billions, cannot be counted or tallied like children through a turnstile, so there must be some other way that satisfies, or comes close to satisfying, the rigors of mathematical formulations without the work of enormously expensive tally counts.

"I have envisioned a thought experiment where there is a chamber with a permeable membrane separating a gas that is cooler from a warmer portion of the same gas. There is some kind of a mechanism that acts like a gate-keeper that judges the molecule according to it's temperature in comparison with the average temperature of all the cool molecules.

"If the temperature of a molecule of cool gas contrasts with the average temperature of all the molecules in a way that lets it cross the membrane, then it will have the effect of cooling the average temperature of the molecules of warm gas.

"Of course the gate-keeper works the same way on the warm side, except it is judging a single molecule in terms of the average temperature of all the molecules on the warm side. One of my general questions is: How would we design a statistical method to help us determine within some range of error how much time must pass until the temperature of the cool side of the membrane must match the temperature of the warm side? What are the attributes of that method?

Eirian's mind had begun to spin. She remembered one of Dodgson's stories in the lecture hall at Oxford two years ago last July, about the Blue Caterpillar sitting atop a mushroom, and how he had become so pontifical. In her mind she could see the caterpillar smoking a hookah, sitting in judgement on the temperature of a particular molecule when compared with the average temperature of all molecules of the same substance. Somehow it seemed an appropriate image.

"It seems almost demonic," thought Eirian to herself.

"Can you craft some methods that integrate changes in temperature and pressure over time for the gases on both sides of the membrane, and then link these changes to the numbers of molecules that must pass the membrane in order to achieve equilibrium on both sides of the membrane?", Maxwell asked Eirian.

====

Taking Eirian through Maxwell's equations…

After taking a break from discussing the movements of molecules in a gas, Eirian asked of Maxwell: "Can you describe your equations of electricity and magnetism to me, and describe what you are trying to measure?"

"Yes, I can," Maxwell replied. "I'll put them on the board for you. Remember that I am profoundly indebted to Michael Faraday for his insights into the workings of "fields," and his thinking about "actions at a distance." Without them, I could not have built the theories that I am about to show you.

"Start by thinking of things that float or move in space without any visible connection to the ground, or to any other nearby physical object. Your imaginary object could be a hot air balloon, an arrow in flight, a cloud, or a puff of smoke. The question is: what force or forces causes each of those objects to change? In the balloon you might think the gradual loss of heat will eventually cause it to collapse. The arrow will eventually fall from the sky… but from what cause? The air it must push against; the speed it will lose with time as it flies?

"How about the cloud? What causes the cloud, or the puff of smoke to change its shape, or it's color? Is it the sun; is it changes in the cloud's moisture content; is it the pressure of the changing wind? Actually, it is probably all of those things."

Maxwell paused to gather his thoughts.

"Think about the clouds again. This time think about the dark, windblown thunderclouds during tornado season in the late spring. Of all the things that are roiling about inside the cloud, what will cause the tornado spout to drop to the ground, or the lightning to come down from the cloud to strike the ground, possibly catching something on fire, or even killing some person or animal. The power of that lightning bolt, the force of it, the heat of it, and the bright flash of light of it can all be measured over time using Newton's calculus, and an extension of that calculus called a "Vector" which has values for both direction and force.

Now think of a bar magnet that you move slowly toward a piece of metal plate. Depending on the size of the magnet, you might not be able to, ultimately, stop it from hitting the metal, and sticking to it in a way that may be stronger than your ability to pull it back.

The power of that magnet, the force that draws it inexorably to the metal plate - like the lightning bolt, but without the flash, the noise, and the drama – can be measured over time using Newton's calculus, and the same Vector extension, but with a math that works for magnets.

"To link an electric force, and a magnetic force to create an electro-magnetic force with a given magnitude, you must use the partial differential equations that I have developed while at King's College.

A normal differential equation looks like this: f(m)dx/dy. This will measure the functional result of a change in m when moved over the distance from dx0/dy0 to dx1/dy1.

"A partial differential equation is used to measure one part (aka one variable) of a multi part function using a normal differential equation. Each of the other parts must use the same method in what is called a chain function. The chain function links all of the partial differential equations (each a normal differential in other words) into a single functional result.

"Mathematicians use a specific symbol to indicate that a function is one part of a multi-part differential, or chain, equation. Instead of d, that symbol looks like this: ∂.

Maxwell smiled at Eirian and winked again. "Is it all much clearer now?"

Without waiting for an answer, Maxwell said: "Now for the hard part."

"Here are two terms I have defined that you will need to fully understand if you are to write any kind of descriptive text about them:

1.) "Div" is short for Divergence. Divergence describes the variance in the space immediately around a point in physical space as specified in whatever inquiry we are making. If Div (the electro-magnetic force) is greater than zero, then Div is more outward diverging than inward; if Div is less than zero, then Div is more inward diverging than outward;

2.) "Curl" measures the tendency of the force to curl, or loop, around the point and gives the direction of the axis about which it curls.

"I have defined the electrical force at a point as the vector E. A vector has both magnitude and direction.

If Div E = 0, there is no electrical force.

"I have defined the magnetic force at a point as the vector H. A vector has both magnitude and direction.

"If Div H = 0, there is no magnetic force.

"If neither Div E or Div H are zero, then an electro-magnetic fore does exist; if either or both are positive, then the EM force will bend outward; if negative, the EM force will bend inward.

"Measuring the Curl requires the application of the partial differential equations needed to specify the change in the positions along the x, y, and z axes over time. These are defined as follows:

"Curl E = -(1/c) $\partial H/\partial t$"

"Curl E can be read as 1 over the constant c (c is defined as a gearing ratio between the E and M force) times the change in H over the x, y, and z axes over time."

"Curl H = (1/c) $\partial E/\partial t$

"Curl H can be read as 1 over the constant c (c is defined as a gearing ratio between the E and M force) times the change in E over the x, y, and z axes over time."

"I'll assume you have got all that," Maxwell said. "I'll emphasize the fundamentals once again… just in case."

Maxwell turned away from the black board toward Eirian, then smiled and winked at her.

"The terms 'div' (short for divergence) and 'curl' are ways of representing how the forces E and H – the EM force - vary in the space immediately surrounding the point of our inquiry. Div is more outwards than inwards (div greater than zero), or more inwards than outwards (div less than zero). Curl, on the other hand measures the tendency of the force to curl, or loop, around the point of inquiry, and gives the direction of the axis about which it curls."

Maxwell sat down in a chair facing Eirian. Eirian's head was down as she quickly tried to capture all the insights into the magic of electro-magnetism that Maxwell had so willingly given her, over this past few minutes.

"I don't know if I will ever understand this," Eirian said, resignedly.

"Don't feel alone in that, Maxwell said. "I don't think that either I, or Faraday have a deep enough understanding of these forces to explain them well even to those of my professional colleagues who believe – following Newton - that the universe is a finely tuned machine. That comforting machine is very much like a Swiss watch; a machine with gears, and winding mechanisms, and hands to show the time, each connected one to the other with no empty space between."

Faraday and I are talking about those spaces… the empty spaces where the power of these forces of electricity and magnetism reside and do their work. To proceed from a lifetime of belief in a clockwork universe to a universe where forces work at a distance with nothing connecting one object to another is hard for anyone to grasp, scientist or not.

"The introduction of these ideas of 'Fields' and 'Action at a distance,' disconnect the forces of the universe from the movements of the objects these forces are thought to influence. Our case will be a very difficult case to make in the face of such thinking, entrenched as it has been over more than 100 years.

"Even so," he continued, "it will be your job to carefully watch and record the details of my experiments with electricity and magnetism. Then, though I won't require it of you, I'd like it best of all, if you will at least try to repeat them without any guidance from me, to see if you can get the same results.

"Whether you try to do your own tests, or not, your final task for me, on each experiment, is the preparation of text and illustrations describing each of the experiments as you saw them. Expect my very careful review because It is in these descriptions that we may all find the best way to present our findings on electro-magnetism to the very large group of my skeptics in the scientific community.

In January Eirian was working at Glenlair. Charles Ludwig Dodgson, whom Eirian and Bethan met at Oxford in 1864, had published a book titled: *Alice's Adventures in Wonderland.* He published the book in November, 1865, using the pen name Lewis Carroll. Maxwell had alerted Eirian to the publication and asked if she would like him to obtain a copy for her the next time he traveled to Edinburgh.

Excitedly, Eirian said "Yes! I'll let Bethan know. She and I can read it together. I'm so excited," she said.

Maxwell then said: "I have an even better idea. I have to go down to Cambridge in March. Would you and Bethan like to come down with Katherine and I? I can make travel arrangements for you both to go to the Oxford campus for a day or two. Perhaps you can meet with Dodgson there, and have him autograph his book for you. Then you can come down to Cambridge to join Katherine and I for dinner with some friends of ours."

Maxwell thought he saw an actual glow in Eirian's eyes. In any case, he noted that she did seem briefly incapable of speaking.

"Oh, thank you, Dr. Maxwell. I am overwhelmed with your kind and generous offer. "Are you sure?" She did not wait for an answer.

"I must go today to discuss this with Bethan. We have so many things to prepare for before we leave.

Two weeks later Maxwell returned from his trip to Edinburgh with four copies of Dodgson/Carroll's book. The first two would, of course, be the property of Eirian and Maxwell. The other two – a surprise to Eirian - would go to Bethan and Katherine. Maxwell told Eirian that she and Bethan would be responsible for getting signatures by the author on his and Katherine's copies.

=====

As with their visit to the several buildings and expansive grounds that make up the beautiful Oxford campus nearly three years before, Eirian and Bethan knew that Dodgson was associated with the school known as Christ Church. Maxwell had alerted Dodgson to the visit by Eirian and

Bethan. Dodgson wrote back to say how delighted he was to be able to see them again. He recalled their earlier visit very fondly.

He said he would be discussing his new novel in one of the Christ Church day rooms the day of their visit. He assured Maxwell that he would have places for them.

====

Before returning to Glenlair, Maxwell, Katherine, Eirian and Bethan are invited to a dinner party in London at the house of Maxwell's long-time mentor and friend, Michael Faraday. The party will honor Benjamin Disraeli. Though not yet resolved, the collapse of the Gladstone government was virtually assured by Gladstone's failure to re-unite the liberals around Prime Minister Russell's Reform Bill following the death of Palmerston. Gladstone, a bitter enemy of Disraeli, would clearly be dispatched from office, along with his liberal party in the upcoming elections of June, 1866.

Disraeli is now 61. He is a friend of Queen Victoria, a novelist, a member of Parliament, and a continuing supporter of Tory governments and politics. He will soon be appointed Chancellor of the Exchequer, and Conservative Leader of the House of Commons in the third minority government to be organized by Lord Derby, a Tory, as soon as he is sworn into office.

====

The dinner takes place at the house of Faraday, Hampton Court, Middlesex. Faraday is now 73 years old. Though without formal education beyond the 8th grade, Faraday has been greatly honored by his scientific peers, and by Queen Victoria herself. He had been offered an opportunity to be buried at Westminster Abbey among the resting places of England's royalty, but he refused, preferring to remain with the common man.

Faraday had been living in Middlesex the past 7 years following his retirement. When told of the visit by Maxwell's young protégé, Eirian, to hear a lecture by Dodgson, Faraday insisted that she invite Dodgson, and whoever else might be good company for him. Charles Babbage, another of Faraday's guests, and like him in that he disapproved of hereditary peerages, though he did favor Life Peerages.

Babbage is now, like Faraday, 73 years old. He has lived at 1 Dorset Street, London, since 1831. Also like Faraday, Babbage, too, was of a liberal and progressive persuasion, though he happily accepted Faraday's invitation to a party honoring the conservative Disraeli.

Actually, this particular get-together was Queen Victoria's idea, communicated to Faraday through Prince Albert – the Queen's husband and the empire's chief science enthusiast - in the hopes of exposing the conservative and sometimes prickly Disraeli to some liberal thinking in advance of the coming change of governments. The choice of other guests at the party, Prince Albert suggested, would be up to Faraday.

Babbage had recently read Dodgson's new novel, written under the pen name Lewis Carroll, *Alice's Adventures in Wonderland.* He expressed his delight at the prospect of his joining the group at dinner.

"Dodgson's novel is the talk of London," Babbage said. His eyes were all aglitter at the thought of Dodgson with his stumbling speech arguing with the articulate Disraeli about the Parliamentary intrigues of the day. When Babbage thought of Disraeli parrying Dodgson's literary wit with some parliamentary and political fantasies that Disraeli would try to promote, Babbage could not restrain himself, and broke into gales of laughter.

When Babbage was told that Dodgson would be accompanied by one of his math students who stuttered, Babbage had to excuse himself and go to another room where he gave in to bouts of uncontrollable laughter and fits of coughing. Other members of the dinner party began to worry about his health, but he returned to their company after a few minutes, waving away any health concerns.

"I apologize to my host, Michael Faraday, for my behavior. I hope you will all forgive me for my ill-mannered behavior regarding the two guests who may not be as fully articulate as others of us. Please trust me when I say that it is not the impediment that I find comical, nor the speakers, both of whom I admire, and consider friends and professional colleagues in Mathematics. The problem is that I imagined the sounds and conversational tension such a debate would carry to the rest of us," Babbage said. "It tickled my funny bone, and my imagination got the better of me."

"Sorry," Babbage concluded.

All nodded their approval of his conciliatory remarks and turned back to their conversations.

Shortly after, an aide to Faraday, acting as butler, announced the arrival of Charles Dodgson and Sean McCabe of Christ Church, Oxford. All rose to congratulate the new author on the publication of his very successful first novel. Though they had all been together on the Oxford campus with Sean over the past two days, Eirian and Bethan approached

Sean, and asked that he join them at the table, when convenient after introductions.

"Th, Th, Thank you," Sean said, stuttering. "I will join you sh, sh, shortly."

In an aside to Bethan on their way back to the table to wait for Sean, Eirian noted that "Babbage would not find the speech problems of the guests so entertaining if we reminded him of his many crusades against minor issues of minimal public interest."

Sean had told them of Babbage being denounced during debate in Parliament two years ago for "Commencing a crusade against the popular game of tip-cat and the trundling of hoops."

The issue was a serious one having to do with interference with traffic along the crowded roads, and with frightening horses pulling carriages in public rights-of-way. But the tabloid press in London had, historically, liked to bring public celebrities down to size with humiliation. They were always looking for targets. They had picked up the issue of 'trundling of hoops,' from the Parliamentary debate, and turned it into a riotous play on language. This had caused Babbage to become a laughing stock.

Eirian and Bethan had waited patiently while Sean explained all this in his halting voice. He was very grateful to the two women for their kind of patience with his speaking.

====

The next day, the Scottish travelers began the three day journey back to Glenlair and Corsock. All agreed that it had been a wonderful trip. The stories they could tell about all the wonderful people they met, and the events they participated in would last them the rest of their lives.

Eirian had been mulling over the ideas in the book "Alice's Adventures in Wonderland," though she was not sure she could see any value in them of benefit to her work for Maxwell. She did love the book, and all of the fantastic imagery and word plays in the book. She resolved to keep thinking about ways to link the magic of the Cheshire Cat, or the Mad Hatter, or the March Hare to the realities of Newtonian math.

For now she turned to Bethan to ask: "You were sitting next to Sean at dinner. Did you notice his speech to have improved during the time he spent with you?"

"Why, yes. Sean's stutter seemed to go away after awhile. I think the stutter is only really there as long as he is under the spell of Dodgson. He is really quite an unusual young man. We will write to each other." "I am delighted to hear it," Eirian said.

For himself, Maxwell, thought about the things he had been told in his business meetings at Cambridge. The University wanted to have a leading role to play in the rapidly emerging sciences of the Industrial Age. The new sciences and the development of new industrial materials and sources of energy were moving fast across all of Europe, and Cambridge wanted to build a science laboratory worthy of that role.

More important to Maxwell, they wanted to hire Maxwell to build it and run it. It would be called the Cavendish Laboratory after the British chemist and physicist Henry Cavendish.

"James Clerk Maxwell will build the lab, and, once built, he will become the first Cavendish Professor of Physics," they said. "…if, of course, he decides to take the job."

But first, Maxwell had an obligation to his beloved father, John, to complete the expansion of the buildings at Glenlair according to John's original architectural vision. Other than work with Eirian to get his scientific investigations and new formulations in calculus needed for deeper investigations into electro-magnetism ready for publication, there would not be much new science at Glenlair during the next several years.

Katherine, Maxwell's wife and indispensable helpmate was not at the meetings, but she could see the far away look in his eyes. She knew that big things were brewing, as they always were with Maxwell, and she would find out about them in due course. She reached over to squeeze his arm.

PANEL 8 - NOW I AM BECOME DEATH

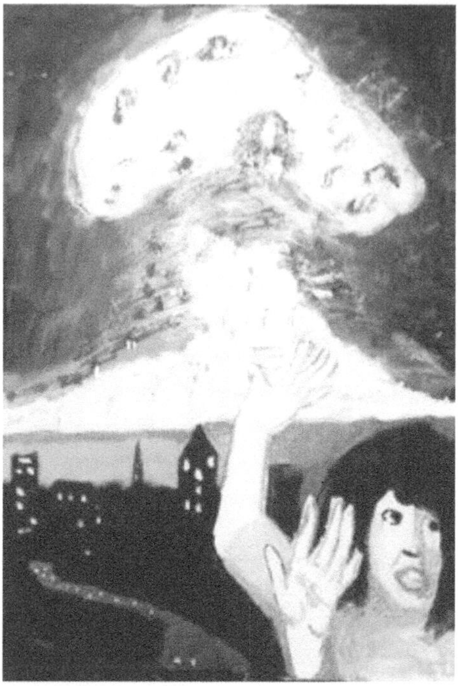

Sarah, a bright young physicist and mathematician, works with Richard
Feynman at a Manhattan Project lab on a mesa near Sante Fe, New Mexico. The site where the lab is located will soon to be named by the owners, the U.S. Army, as Los Alamos. She writes the setup procedures for tests of mathematical formulations needed to control shapes and expansion rates for explosion blast effects.

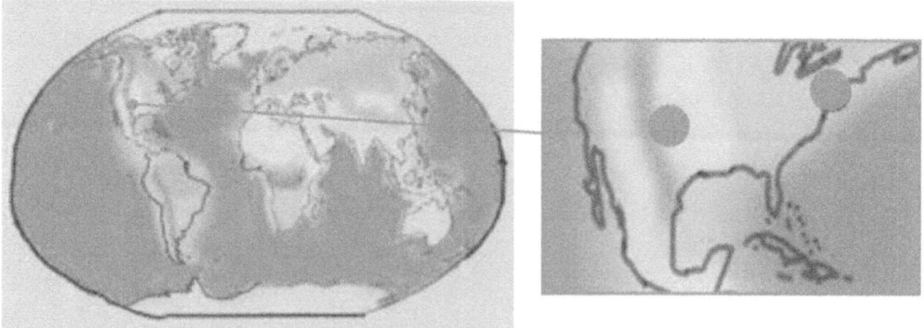

LOS ALAMOS AND NEW YORK CITY. FALL 1944 CE

PROTAGONIST

Sarah - A bright young woman in her late 20s who has a lot of talent in mathematics and physics, but her official job involves helping with the documentation of the theoretical design and testing work needed to make an atomic bomb.

Konstanty - Co-worker with Sarah; responsible for setting up and managing the explosive testing of bomb configurations built according to calculations done by Sarah.

Meuric - Co-worker with Sarah in supervising some of the "Computers"; actually the men and women who did manual calculations in mathematics with data generated by the scientific work in the Theoretical Division.

Thomas - Recruited by Sarah to help manage the development of the mathematics needed for bomb design based on the results of field testing of possible bomb configurations, as calculated by the project's human "computers.".

ANTAGONIST(S)

Richard Feynman – (b. 1918 CE – d. 1988 CE) a brilliant young physicist only 24 years old with a brand-new PhD when hired for the Manhattan Project. In addition to Feynman's early accomplishments in nuclear physics, he was also – from childhood - very skilled in trouble-shooting and fixing household, automotive, radio, sewing machine, and small business mechanisms and systems.

LESSER CHARACTERS

Dorothy McKibbin – (b. 1897 CE – d. 1985 CE) Executive Assistant to Robert Oppenheimer. Responsible for the project compound at 109 East Palace Drive in Sante Fe. Dorothy became the first point of contact for every member of the Los Alamos Staff when they first arrived to begin working on the Manhattan Project.

MINOR CHARACTERS

Manhattan Project Leaders at Los Alamos

Leslie Groves – (b. 1896 CE – d. 1970 CE) Leader of the Manhattan Project. An Army Brigadier General appointed by FDR in September of 1942 to rapidly build and manage the sites, facilities, systems, and the military and political interfaces needed for bomb design and development.

Robert Oppenheimer – (b. 1904 CE – d. 1967 CE) Leader of the Science Program of the Manhattan Project. Appointed by Grove; answers primarily to Grove on project issues outside of the science.

Hans Bethe – (b. 1906 CE– d. 2005 CE) Leader of the Theoretical Division assigned to recruit the scientific staff, and to oversee the design, development, and maintenance of all the sites, laboratories, and other facilities and testing environments needed to determine and resolve theoretical problems. Direct supervisor of Richard Feynman, who, in turn, supervises Sarah.

It is September, 1939. Albert Einstein represents theoretical physicists working to develop methods for splitting the atomic nucleus and controlling the chain reaction that otherwise leads to an atomic fission

explosion. He writes a letter to U.S. President Franklin Delano Roosevelt expressing their collective fears about German scientific progress in nuclear physics. Adolph Hitler, self-proclaimed "Leader" of Germany's reinvigorated industrial base and military forces could be trying to build an atomic bomb. In concept, such a bomb would further enhance Hitler's war-making abilities by frightening any nation resisting his efforts to take their lands by military invasion.

With such a bomb, he could destroy the evolving coalition – in late 1939 - of all states and territories held by Hitler's primary enemies, Britain, Russia, and the United States. Responding to Einstein, FDR authorizes the Manhattan Project. By 1944, 130,000 people are involved in the several Manhattan Project locations across the U.S. All were dedicated to the top-secret design, testing, and accelerated production of a deliverable atomic fission bomb.

The theoretical design, development, testing, and prototyping work on the bomb takes place in the mountains on the site of a former dude ranch on a mesa near Sante Fe, New Mexico. With the U.S. Army as developer under Pentagon builder General Leslie Groves, Los Alamos sprouts like an old west gold rush boom town.

One of the most difficult initial problems in the theoretical division involves the work to refine the multivariate differential calculus developed by James Clerk Maxwell 80 years before in his work on electromagnetism. Maxwell's equations must be refined into the narrow focus needed to predict, precisely, the size and shape of the two lumps of radioactive material that must be forcibly merged to create the "critical mass" needed to initiate the fission reaction. The equations must further enable the precise design of the size and shape of the chamber that would contain the non- nuclear merging process; thus creating critical mass, and the atomic fission explosion that will then unfold rapidly within a few millionths, or billionths of a second.

Some scientists were sure that there was at least a theoretical

chance that such a nuclear explosion could catch the Earth's atmosphere on fire. Sarah, a bright young mathematician was brought in to work with Richard Feynman's "Diffusion Unit" to design the shape of the bomb's implosion chamber. On deep background among managers, their careful work – it was hoped - would resolve most of the fears of the air-on-fire people.

Konstanty had called out to Sarah. "These calculations are fucked up, Sarah! Do them again."

Later, back at work after a visit to the hospital where Konstanty was in a coma, Sarah remembers swearing under her breath. She and Konstanty held the same status in Feynman's lab, but Konstanty seemed to think he was smarter than her, or had more pull with the project's science managers, Oppenheimer, Bethe, and Feynman. For the sake of office peace, however, Sarah had picked up her notes and read, once again, Dr. Feynman's brief scribbled notes on the problem.

Feynman had never really sat down with Sarah to discuss the overall purpose of his unit in Hans Bethe's Theoretical Division, nor his personal, particular, and specific, interest in the work at hand. Feynman's skills, experience, and sheer brain power was so beyond anybody in Sarah's experience that he was often called out to work on various technical emergencies on the Project. Clearly, though, as one of the world's leading physicist – certainly the leading physicist in the world for his 24 years of age and his very green, two year old PhD – in the area of quantum physics, quantum mathematics, and quantum mechanics, Feynman's work on the Manhattan Project, from the earliest days with Fermi in Chicago, had always been totally clear, and totally focused.

So far, in her only face-to-face conversation with Feynman, he had pointed to an Organizational Chart for the Project's Theoretical Division, and to a box fairly high-up on the chart labeled "Diffusion Unit."

"My official job is Group Leader for the Diffusion Unit. Among other things, I am responsible for the mathematics, and the design, configuration and testing of various ideas about the size and shape of explosive effects; how to manage the effects of bomb blasts in other words. Your job is the mathematics. Let me know if you have any questions."

Below the Diffusion Unit box were branches to two smaller boxes. One was marked "Computation", and the other was marked "Field Services." "Computation" represented a small staff of mostly women – often the wives of project scientists and staff - who would work through the masses of data generated by a single test. "Field Services" represented a small staff of mostly men who would build the blast chambers to Sarah's specifications, and then they would actually build and install the explosive devices, and the firing and safety mechanisms needed to control the explosion, and the sensors and telemetry needed to generate data on the composition of blast effects.

Finally, once everything had been installed and tested for normative function, the Field Services staff would get to touch off the explosives and create a big boom, usually with lots of smoke. It was a great job, though dangerous as hell, as Konstanty eventually found out to his disadvantage.

As best Sarah could understand it, then, her job in the unit had been to support the development of a calculus for expansion rates of gasses and particles that have been blown up by an explosion. From scuttlebutt among some people she knew around the project, Sarah had gleaned some insight into why the precise description of exactly how the particles at the subatomic level would begin to merge once the fissile material reaches critical mass.

Once the nuclear fission reaction began, the force of the explosion propagated through the space containing the explosive material was also critical. the shape of the container that would hold the explosive material - itself shaped - would dictate the force, direction, and effectiveness of the explosion.

In one design, known as the implosion model, two pieces of highly radioactive plutonium are jammed together in a precisely shaped chamber by a conventional explosion. The efficiency of that jamming would also dictate the efficiency of the resulting nuclear fission, and, therefore, the success of the bomb itself. In a second design, known as the gun model, a piece of highly radioactive Uranium 235 is literally shot at a second piece of highly radioactive Uranium 235.

Four bomb design formats based on the likely choice of bomb fuel between uranium and plutonium had been adopted by the project's top level political, military, and scientific managers. These top level managers had to make a determination on the costs of producing both types of fuels when compared with the efficacy of bomb design, development, and delivery. These factors could not be known until the work of the Theoretical Division in testing various bomb design prototypes was well underway.

To a large extent Sarah had been on her own in the development of the precision calculus, and the physical testing of various bomb and container shapes. She had learned early in the process that, advanced calculus using multiple partial differential equations is fairly straight-forward, but equations that must account for quantum factors and must use statistical mechanics in reporting the resulting output is almost infinitely more complicated.

The word quantum had not yet entered the scientific vocabulary, but Isaac Newton had called the gravity that holds planets in orbit "action at a distance." Einstein did have the word quantum available, but he did not accept the implications of it. He called the quantum measurements that hold the atomic nucleus together "spooky action at a distance" -

Later, as Sarah set up her office, she left a note for Feynman asking his advice on reference materials for the latest thinking on the mathematics of quantum physics, he replied with another scribbled note: "If you don't know 'Maxwell's Equations,' start there, and write me a summary of what you learn." Sarah was momentarily puzzled. Surely thinking that was 80 years old – as Maxwell's was – could not be thought of as "...the latest thinking" in the rapidly evolving science of nuclear fission in wartime. Nevertheless, she resolved to find the paper Maxwell had written and published on the subject in 1864 titled: *A Dynamical Theory of the Electromagnetic Field.*

Eventually she came to understand that detailed design on the prototype bomb itself would depend on the accuracy, perfection, and reliability in the results predicted by whatever calculus Sarah would have come up with.

For all his bull-headedness and bad-mouthing, Konstanty had helped her in her work. Also, he had been right about the calculations derived from the particular test configuration that got his hackles up. A week ago, unfortunately, Konstanty had been seriously injured by a test explosion gone wrong. Sarah had been devastated, but the pressure of the work was so intense that she had to put her feelings for Konstanty aside as she began a search for another assistant; one who had at least some ability to help guide the human "computers" doing the computations by hand that would eventually lead to something much more mysterious, and much more dangerous, than mere analysis.

Thomas is a recent Master's degree graduate in Math from Brown University. He had solid paper credentials for the work, had published a couple of articles on the application of Statistical Mechanics, and could pass the intense security surrounding anything to do with the project. Also, he was willing to put up with the highly secretive, and sometimes aggressive recruitment process that preceded any appointment to any scientific job on the project. Sometimes that willingness, by itself, raised red flags among those involved with managing recruitment, and the resolution of these concerns usually required an extra push by Feynman, by his boss, Hans Bethe, and by the Project Science Manager, Robert Oppenheimer.

On the project, the most difficult security problems came up in recruiting and hiring of specific scientific talent. Some recruits came from countries now at war with the US; some came with very questionable associations; and some came with both. When necessary, Oppenheimer might have to ask the overall project leader, Brigadier General Leslie Groves, to get involved, and Groves, when absolutely necessary, had direct access to FDR himself.

Though Groves avoided the possibility of it like the plague, he would, if it became absolutely necessary, ask FDR to personally intervene. However, hiring an unknown mathematician like Thomas for a technical job that was a few levels down the org chart did not rise anywhere near that level.

Thomas's main problem, it seemed to Sarah, was that the job candidate is young, and does not know anybody in the higher reaches of the kind of advanced Mathematics and very new Physics in play on the project. After all, Enrico Fermi, the Italian Physicist, formerly a Professor of Physics at the University of Florence, had succeeded in conducting a sustained and controlled nuclear fission reaction in the basement under an unused squash court at the University of Chicago on December 2, 1942; …only about 6 months ago. Plutonium, along with Uranium, one of the two materials to be developed and tested with designs developed by Bethe, Feynman, and the physicists on the staff of the Theoretical Division – possibly to include Thomas - for the most efficient bomb fuel, only existed in theory, and not yet been made to exist in physical space.

As a side note, Fermi himself, with his Jewish wife, Laura, was a live example of what was really at stake in the Allies' global war with the Axis powers of Germany, Italy, and Japan. No Italian scientist, especially one as accomplished as Fermi, was allowed to leave Italy under the Fascist regime of Benito Mussolini. But Fermi's early work in nuclear physics was so fundamental, and so important that he was awarded the Nobel Prize for Science in 1938. The Italian government allowed Fermi and his wife to go to Sweden to accept the prize. After the ceremony, Fermi and his wife never again returned to Italy.

On the day Fermi's "atomic pile" in Chicago created and had sustained the world's first nuclear fission reaction using Uranium 235 the coded message to FDR was: "The Italian navigator has landed in the New World."

But, according to his advisors and professors, Thomas has studied James Clerk Maxwell's primary, world-changing work on electro-magnetism titled:" The Dynamical Theory of the Electromagnetic Field," published in 1865. He had written a couple of papers extending some of Maxwell's ideas in ways that could enhance the work in the Theoretical Division. Plus, in the words of one mentor, "There is no question in anybody's mind that Thomas is the smartest guy in the room."

"Even at his young age," one of his advisors said, "Thomas has written at least two very insightful papers on Maxwell's experimental work on optics, as well as additional monographs on electricity and magnetism, and he understands the fundamentals of math in new and creative ways."

It is these heartfelt recommendations that caused Sarah to literally ignore all the other noise associated with the somewhat off-beat upbringing and family of Thomas.

In the case of Thomas, Sarah, now top operational manager under Feynman while Konstanty remained under intensive care in a hospital, had such an urgent need for someone who knew enough advanced math to oversee the calculation process, that she practically begged Feynman to speak for Thomas with Hans Bethe, and with Oppenheimer if it became necessary to ease Thomas past any security concerns.

Thomas had one other problem, interesting, and largely unprecedented since Alexander Hamilton fought alongside General George Washington in the Revolutionary War; Sarah told Feynman that: "Thomas has family in Haiti, and, though his parents now live in New York City, one of his parents is black."

Aside from his brilliant work in theories relevant to bomb design, Feynman was a very popular guy around the project mostly because he was always available to help fix some kind of machine or pick a jammed lock that was holding up progress on some other scientist's urgent project. He was also a fun-loving guy who enjoyed playing practical jokes on his co-workers; he enjoyed leading dance contests in the several after hours clubs scattered around the Mesa in various private homes, and he played the bongo drums.

So… Feynman was initially reluctant to put a word in for Thomas. From Sarah's description it sounded to Feynman that he might have to use up many, many of his precious political and organizational chits in order to bring a half-black untested mathematician with a Caribbean accent onto the project. Worse, in Feynman's eyes, Thomas may not really know anything useful about the new mathematical minutiae needed to keep the project's fission design, testing and development processes on track.

Sarah continued to be persistent, insistent, and desperate enough, though, that Feynman decided to include a meeting with Thomas during his next trip to the east coast to consult with project scientists working in secure and anonymous facilities on the University campuses of Columbia, Cornell, Princeton, and MIT.

Sarah, delighted with Feynman's initiative on her behalf made the arrangements by phone after working out the details with Grove's designated security officer on the Mesa. Thomas would meet with Sarah's "Departmental Superior" at the University she was "…recruiting for." When Thomas asked what University that might be, Sarah could only say that they were involved with a wartime project that was top secret, and under fairly tight security. Thomas accepted this idea and agreed to wait for the "…Departmental Superior…" who would simply sit down at his table at a well-known restaurant in mid-town Manhattan and introduce himself as John Brown.

The "well-known restaurant" and the details of when and which table, would be made known to Thomas within two days of the meeting. His travel and hotel arrangements for the meeting would be taken care of. He would only need to come to the designated ticket window at the railroad station in downtown Providence, Rhode Island to pick up his paid-for ticket package.

In the back of Thomas's mind, he could not entirely suppress the idea that this was all a huge joke. He had an imaginative turn-of-mind. If he let his imagination flow freely, he could see someone in a plain, dark brown overcoat jumping out of the bushes at him, threatening him with a handgun that popped a flag out of the barrel marked BANG! He chuckled at the notion.

=====

Sometimes, Sarah would get bogged down in a tangled pursuit of solutions to one of the problems typical of Feynman's many task requests. During her advanced courses in calculus, she had learned to recognize the point in the pursuit of a solution when she needed stop trying to force a path to an answer and take a mental break. She would get up from her work table, and walk around, or she would lay down on the couch and take a nap, or she would lean back in her chair and begin a process of wool-gathering over things that she knew to be more pleasant to deal with.

She remembered the day – Friday, April 30, 1943 - she arrived at the train station in Lamy, New Mexico, after a three-day trip from Chicago. Lamy, a tiny town with a big railway station, and a big nearby hotel, the El Ortiz, was as close as the Atchison, Topeka, and the Sante Fe railroad, ever got to New Mexico's state capital at Sante Fe. Engineers who laid out the original route of the railroad realized that the hills around Sante Fe made it impossible to run the railway directly into the city.

A small spur line had been extended from Lamy to Sante Fe in 1896, and many of the new arrivals took this rail spur the 14 miles from Lamy into the small station in Sante Fe. With Sarah, however, Feynman wanted to make sure she got the best possible treatment on her way to the hectic, gritty, muddy world on the Army base and the barracks style buildings known to the civilian and military inhabitants as Los Alamos.

Sarah was approached by a civilian on the station platform, who greeted her by her first name. She did not know the driver, but she had been thoroughly briefed on the security procedures on the project. She knew a strange man in civilian clothes would approach her, speak her first name and take her bag to a car parked nearby. Once in the car, Sarah was then driven to a pleasant residential compound at 109 East Palace Drive near downtown Santa Fe. The entry was marked by a small, blue sign with red lettering that read, enigmatically: U.S. ENG—on the top line, and then, on the bottom: RS. This little sign was the only sign allowed to indicate the only point of contact in the city for almost all the people, and much of the material passing through Sante Fe on the way to Los Alamos.

Once inside the compound, Sarah was taken to a small office in the back. There, Dorothy McKibbin greeted her warmly.

"I think you must be Sarah," Dorothy said, with a smile. "Welcome to Sante Fe."

"Thank you," Sarah responded.

"You will have to fill out this form so I can give you a letter of authority to go to Los Alamos."

Sarah had just come out of the warm and rather carefree embrace of academia. She was not used to the possible demand for her official "papers." Jokingly, she asked "What happens if I don't fill out the form?"

"Then I can't issue your pass," Dorothy responded without smiling. "I'll let you think about what might happen on a top secret military base if you are caught without documented authorization for your right to be there." Sarah, admonished, took the form and carefully filled it out.

====

Dorothy McKibbin was a widow with a 12 year old son in March 1943, and she was out of work.

Dorothy Scarritt was a 1919 graduate of Smith College, located in

Northampton, Massachusetts. When growing up in a well-to-do Kansas City, Missouri family she was well-known and well-liked. It was expected that she would have a bright future in whatever she chose to do. Unfortunately, she had not had the kind of luck in her life that would give her a stable and comfortable living. Originally born and raised in Kansas, Dorothy Scarritt's academic record in school was outstanding. She had left for Smith College with the best possible start, and with continuing high achievement through college toward a good life.

After graduation from Smith, she began, as she had been raised and educated to do by her family, to seek her fortune. She had not found fortune but had found the man she wanted to marry. Then she came down with Tuberculosis. In that time before anti-biotics, the high, clear, desert air in the American southwest was thought to be the best way to pass the time while hoping for a "cure" for this incurable disease.

Dorothy's mother got her into a sanatorium in Sante Fe. Dorothy broke it off with her boy friend Joe McKibbin, because she felt it unfair to him to be asked to wait for the years it might take until her tuberculosis resolved itself one way or the other.

Then, by a major stroke of luck, Dorothy Scarritt recovered from tuberculosis. She was declared "cured" in 1926. Joe McKibbin re-committed to her, they got married, and had a son named Kevin. They set up housekeeping in Sante Fe. Dorothy loved the ambiance of the place, the relatively slow pace of things compared with her days growing up in Kansas City, and she loved the diverse and abundant arts community.

Unfortunately, Dorothy's irregular runs of luck went against her again when Joe died of leukemia shortly after their marriage. Dorothy didn't have much time to grieve. Without a major bread-winner she would have to find work. Fortunately for her she found a bookkeeping job with a local retail company in Sante Fe, the Spanish and Indian Trading Company. That job lasted 10 years. Unfortunately, again, for Dorothy, the bombing of Pearl Harbor, and the beginning of the war changed the focus in the Sante Fe community from tourism and trading to the departure of many of the young men to help the war effort. The Spanish and Indian Trading Company had to close its' doors.

Then, while unemployed and looking for work, Dorothy – by another major stroke of luck – got a tip from an enterprising local businessman that a large organization was planning to build a big facility nearby. They needed somebody who had deep office management skills; was both intelligent and very familiar with the area, and they needed that person to start work as a Secretary right away.

"I need to know your answer in 24 hours," said the local – an entrepreneur named Joe Stevenson - who gave her the tip.

Dorothy thought to herself that Joe had been away in California for several months, and he seemed much more secretive now. He used to be so fun-loving and personally engaging. Now… much less so.

Dorothy had, that very day, been offered a job as a typist in a local bank. The bank job would be well understood, simple, and there was virtually no chance that the job would ever go away, as had her 10 year job doing the books at the Spanish and Indian Trading Company. That job went away because so many of the men and women in Sante Fe were leaving to go to war, and many small businesses were closing because they were losing key employees. Demand for local goods and services was also collapsing by reason of the shrinking population.

On the other hand, the Secretary job was not well-defined, nor even barely defined by the man who invited her, with some urgency, to think about taking the job. All she got from the entrepreneur was that it was a government project and might be important. When Dorothy asked any kind of penetrating question about the job, all she got was downcast eyes, followed by a blank stare.

Her ambiguity on the question of which job to take disappeared when she met the man who she would be a Secretary for. The "job interview" occurred in the lobby of the very posh, very old La Fonda Hotel. The man Dorothy had been told she might be Secretary to, was a tall, attractive man with an intense look. His eyes did not look away from her when he was introduced to her. He wore a porkpie hat tilted at a rakish angle. The man exuded so much confidence that Dorothy knew that the job they might offer to her would be important. She had found the job she wanted. The job as Secretary had at least two added benefits: It paid $150 per month; much, much better than the job as a Sante Fe bank clerk; and, second, she would be working in an office in town.

She had feared that she might have to work in some remote, isolated site away from the life, the amenities, and the comforts that she valued in the small, artsy community of Sante Fe.

Oppenheimer had said virtually nothing to Dorothy beyond some small talk about the weather, and some local artists. After a few minutes, Oppenheimer is reputed to have said: "All right. I'll leave you then."

After Oppenheimer left the lobby Dorothy quickly found Stevenson, grabbed his arm, and announced to him "I'm in."

====

One day late in the summer of 1943, Feynman popped into Sarah's space and admonished her, "You need to be constantly reading about the evolution of knowledge about the structure of the atom, it's nucleus and its cloud of electrons, since Faraday and Maxwell," he said.

"More important even than that is the knowledge gained about the interaction of the electrons in the atomic cloud with the atomic nucleus since Einstein wrote his *General Theory of Relativity.* That is the source of all practical knowledge under quantum theory, and the only way to get at the distributive detail you need to calculate the details of the collisions with other particles, and the follow-on collisions that will result," Feynman continued.

"We cannot model the shape of the bomb blast, nor estimate its' effects downrange without that knowledge. Got it? Use whatever notes you can find in my office to get the knowledge you need."

Feynman departed to go on to his next task. "He always," Sarah thought, "has so many tasks." Sarah went into Feynman's library to search through his books.

"Feynman popped back into the room for a moment. "You could start

with my notes on Fermi's work on the pile in Chicago for an overview of how I view these interactions and how they are driving the work here. My Doctoral thesis might also be useful in understanding my work on 'sum-over-paths.'" Then Feynman was gone, and Sarah had no idea when she might see him again to ask questions about the things she would be reading.

====

By late summer, calculating speed and power in processing the huge quantities of data emerging from testing designs and configurations, has become the single biggest bottle-neck in the work of the Theoretical Division. Each configuration test run against protocol produces hundreds of thousands of data points; all had to be processed by hundreds of people called either "Computers", or "Calculators" using hand crank powered mechanical calculating machines sometimes called "comptometers," and built mostly by Marchant.

International Business Machines – commonly referred to as "IBM" – had been in the business of providing machine-based adding and subtracting for commercial transactions since the 1920's. Though well-established, IBM machines did minimal processing beyond fast and accurate arithmetic adding and subtracting. With some technical trickery, their machines might be able to multiply sales in a month by 12 to get an annual forecast, but had almost no calculating power beyond that, and no presence in the scientific market otherwise.

Processing the data on the manual Marchant machines required a series of steps in complicated but rigorous sequences. A single dataset often had to be entered into one of the machines, processed to a result then tested to make sure the data in the result was "normative." In other words, the processed data was within expected ranges. If the tests for "normative" were satisfied then the data in the result could be entered into a different machine with a different set of steps, and a different set of normative parameters.

These complicated sequences with repeated entry and re-entry of processed and re-processed data was prone to a lot of errors. Results had to be constantly checked and re-checked. The constant pressure to do careful, detailed work, and to produce accurate, verifiable results was intense. Work shifts were limited to four hours. Otherwise the human computers wore out too fast, the error counts started to escalate and the process downtime started to accumulate to unacceptable levels. Some of the staff started asking for transfers to something different in some other unit of the project.

Feynman's talent for fixing broken machinery was constantly on call by the "computers"; ...i.e., the staff dedicated to the data entry, and the verification and validation of the computation results. He responded quickly because the results produced by the computers were critical to the overall performance of the Diffusion Unit. Feynman knew he couldn't allow this backlog situation to continue, if for no other reason than Bethe felt the machine maintenance was taking away too much of Feynman's time when it was actually needed for work on the theory, design, and development of the bomb itself.

Eventually, Bethe, for the very reason of his concern about the allocation of Feynman's time, told Feynman to stop doing machine maintenance. Feynman's valuable time and talent was needed on theory, not mechanism. In a way Bethe's concern mirrored the general concern across physics since Newton, Faraday, Maxwell, and Einstein: "Is there such a thing as a 'real' object, with mass, dimensions, and boundaries that can be measured by numbers that can be added, subtracted, and in other ways 'calculated,' or not?"

Making progress in finding an answer to that profound and persistent post-Einsteinian question was very much within the mandate of the Theoretical

Division, and, therefore, very much within the direct and explicit responsibilities of Bethe and Feynman. The problem in the way of making progress on that question was the severe under development of raw calculating capacity in science. Early in the 19th century, George Babbage and his protégé, Ada Lovelace, with extensive financial support by the British government under Queen Victoria and her husband, science enthusiast Prince Albert, made some great strides in developing the actual machinery and programming techniques for doing scientific calculations.

These were to be working machines driven by steam. Unfortunately, Babbage was only able to build a small working prototype before the escalating costs of development began to approach that of a major warship for the British Navy. The main problem seemed to be that lack of well-developed standards of manufacturing for metal, particularly the brass used by Babbage, meant that the lack of best practice standards in developing materials and procedures used in machining things like bolts, nuts, and screws caused errors that accumulated and eventually stopped a complicated calculation procedure before it could produce a result.

The Manhattan Project managers reasoned that, as the largest and most well-known of the providers of commercial data processing services, IBM should be able to help with the growing calculation problem.

====

The calculation problem at Los Alamos began to grow rapidly because of the lack of knowledge about the nature of interactions at the atomic level of matter. There had not been the need, nor the political will, thanks to the experience with Babbage's Difference Engine, to provide the money necessary to explore - beyond paper and pencil in the classroom and university laboratory - all the questions that surrounded the exploration of nuclear energy at the subatomic level.

That casual thinking all started to shift when Adolph Hitler was able to maneuver himself into the leadership of the German government in 1932. By the time Hitler's Wehrmacht invaded Poland using blitzkrieg (coordinated, fast-moving, air attack, tank movements on the ground, and close infantry) tactics in early September 1939, the need for exploration shifted into a white heat with Einstein's letter to Roosevelt.

Unfortunately, the problem in the way of finding answers to nuclear fission questions was not only about the money. It was also about the application of applied mathematics in ways that went far beyond anything that had ever been tried before. The calculus developed by Newton in the 18th century could provide rigorous approaches to many of the known nuclear forces that needed measurement, but there were so many unknown aspects of the interactions of those forces that could only be understood by brute calculations of billions of nuclear particles moving among each other at speeds approaching light speeds with all their accumulated and explosively expanding masses.

In the explosion contemplated for the bomb under design at Los Alamos, the time between achieving critical mass in the explosively merged chunks of plutonium fuel, and the maximum explosive yield was a matter of a few billionths of a second. A few billionths too early and the fission would stop without exploding; a few billions too late and the effects would be unpredictable, and possibly catastrophic.

The factors affecting the length of the time frame between a failed reaction and a possibly catastrophic one includes the mass and shape of each of the two chunks of plutonium to be merged; the shape of the merged chunks at critical mass; the shape of the container containing the merged chunks; and the purity of each of the materials in the explosion chamber.

IBM machines with their minimal ability to do simple multiplication of numbers with two decimal places or less were not going to get the job done. If IBM managers and engineers could be persuaded to study the problem carefully enough to understand the scope of it – and recognize the possible scope of the increased revenue success might produce for their company, then Feynman figured he might have a chance to solve the calculation problem that Bethe wanted him to solve; ...and the sooner the better.

In the meantime, the science of the atom, and the design, development, and testing of the nuclear fission process must move forward. Feynman would have to make do with human calculators doing manual data entry with the clunky, noisy, clanky, and aging Marchant machines. At the same time, another group of humans would have to do the kind of manual verification of columns and columns of result data that would be sure to make normally sane and cautious men and women go a little wacky with too much concentration and focus.

=====

Most of the computers were women. They were wives of scientists and other civilian and military staff on the secret military reservation that is Los Alamos. Some of the wives were educated in Mathematics or Physics, one or two had Ph.Ds in their field, but the only work available to them given their education, training, and experience was calculation and manual data processing. Whatever their training for the work, they were only paid 3/8ths of a normal salary, because they could only effectively do the concentrated work for 4 hours a day.

Altogether there were nearly 100 woman employed on the project at Los Alamos. Several had other jobs or assignments, but most, at one time or another did calculations, at least until the IBM machines could be integrated into the process toward the end of 1943. On any given day probably 50 or 60 women, with some men, could be found at work doing calculations related to the trajectories, speeds, and mass of each of the atomic particles using Marchant machines, paper and pencil, and a variety of processing tips, tricks, and "work-arounds" suggested by Richard Feynman.

=====

The problems to be solved involved a common physics problem called Diffusion. Early in his career, James Clerk Maxwell, using a thought experiment, tried to understand how the molecules of a perfume inside a bottle, once opened to the air, would diffuse until they, and the smell they carried, had filled a room. He envisioned a container with a measured amount of warm liquid in one end, and the same measure of a cool liquid in the other. A membrane separates the two, but there is an opening that allows certain molecules of the warm liquid to escape to the cool side, and, vice-versa, allows certain molecules on the cool side to escape to the warm side.

Maxwell's question had to do with what kind of a mechanism measured the speed of particular molecules and compared that speed to the average speed of all molecules in either the warm liquid side, or the cool liquid side. if the particular molecule was fast enough or slow enough, or met some other test then the molecule could pass the mechanism's gate-keeper function to the other side of the membrane.

Maxwell wanted to understand this decision-making mechanism – it came to be known as Maxwell's Demon - because that mechanism was the only way to come up with a precise answer to the question "How long until the molecules in both sides of the container are exactly the same temperature?"

Maxwell's attempt to measure the flow of molecules in warm water did not compare with the explorations and tracking of particles in the atomic nucleus reacting to a fission process where billions of particles are moving at near light speeds. The fact is that the problem of tracking atomic particles through a diffusion process at the subatomic level cannot be solved by the existing computer technology of the present day.

Computing machinery and programming with the scope and design needed for the magnitude of calculation needed by the Project had been shown to be feasible in a crank-powered prototype nearly 100 years earlier by Babbage and Lovelace.

Unfortunately for the managers of the Manhattan Project, the technology that needed to be developed to extend the Babbage-Lovelace prototype into a production capable calculation system had never happened. Whatever happened when Babbage and Lovelace demonstrated their prototype to Queen Victoria, and her husband, as well as England's science and math enthusiast and promoter, Prince Albert, no further development money would ever come to Babbage-Lovelace from the British government.

The Project's military and scientific managers had no real choice but to create a formal Diffusion Unit and put Feynman in charge of it. It was here where Bethe's choice of Feynman to be group leader over the Diffusion Unit instead of someone like Edward Teller, a Hungarian scientist, would begin to pay big dividends. Teller had felt himself to be entitled by his professional achievements and accomplishments in the study of nuclear physics, to the job that Oppenheimer had given to Bethe. Teller had lingering hard feelings about the treatment he had received from Oppenheimer but had prepared himself to take the job that Bethe had now given to Feynman.

Feynman's intelligence, energy, sheer brilliance, and - best of all – his long experience with fixing radios, and the other emerging electric gizmos of the world, in Bethe's mind, made Feynman extremely well-qualified for the job. By the time he came to work at Los Alamos, Feynman knew in his gut what caused them to break, and he knew instinctively how to either fix them, or work around the busted parts, or restore them, or the parts of them, to some measure of functionality, or to salvage them for some other purpose in the larger system.

Feynman was also the guy who would try anything, including the dumbest and nuttiest sounding ideas, even when the opposition to his attempts were hostile, in his attempt to try to make something not working into something working. The result of Feynman's wildly creative, but often scatter shot approach to problem solving often succeeded where others had given up long before. There was no one better qualified for taking on the diffusion calculation problem than Richard Feynman.

====

Feynman met with Thomas at a small restaurant called "Julios' Bistro – Bar and Grill" on the corner of 50th Street and First Avenue in mid-town. He recognized Thomas from the pictures Sarah had shown him. He walked up to Thomas, sat down, and announced: "Hello Thomas. My name is John Brown."

Thomas noted that John Brown wasn't much older than he was. "Should I shake hands with you?", Thomas asked.

"Of course," Feynman replied and stuck out his hand. "We're pretty secure here, but I can't tell you exactly how secure. If you come to work for us you will begin to understand the full extent of it. I understand you are from the Caribbean, and you know something about math. Is that right?"

"My father was born and raised on Haiti. His parents – my grandparents - were both black at least for a few generations back. Few people in the Caribbean will hazard a guess about bloodlines if their lineage has origins among slaves working the plantations. Whatever the bloodlines, my father was an accountant, and he was successful enough to be recruited by a nationally known US accounting firm, then relocated to New York, where he met and married my mom.

"My mom was a math teacher, and she and my father both spent a lot of time with me, on my homework. Now, I like to think that I know quite a lot about mathematics, especially the application of some applied math in the area of statistics that has never been attempted before. I don't know what you can tell me, but whatever you can tell me about your current problems in computation, I can certainly tell you what I know, and have experience with."

Feynman paused. "I've read your papers on Maxwell's work in optics and electro-magnetism. We are finding that some of the mysteries he and Faraday exposed at the molecular level are still mysteries today, but they are mysteries now more at the level of the atomic nucleus and surrounding cloud of atoms. As you probably know, the tools of investigation even at the molecular level are not much more developed now than they were for Maxwell.

"I agree with what you say about Maxwell and Faraday, but what can you tell me about the math and computation problems you are having at the subatomic level?", Thomas asked.

Feynman stared at Thomas for several moments. Thomas was calm under this close scrutiny. "Do you know any magic tricks?", Feynman asked.

Now it was Thomas' turn to stare at Feynman. He stared for about the same length of time as Feynman had stared at him. "Is this a question about whether or not I believe in black magic; whether or not I believe in the scientific method; whether or not I might prefer to believe in Voo Doo over mathematics?"

Thomas frowned as though he had been profoundly offended. Then he looked up at Feynman to see if he could read his intent. Feynman, a master poker-player among his many talents was both expert and well-practiced in hiding his "tells." There were many poker games going on at Los Alamos, and Feynman was a player in a few games every week.

"It is a simple question," Feynman said. "Give me your honest answer so we can move on."

Thomas thought for a moment then raised both his hands in front of his face and waggled his fingers at Feynman. "Booga, Booga," Thomas said.

Feynman's bursts of raucous laughter could be heard across the room. This went on for a few moments until Feynman was able to get control of himself. Several heads in the room had turned in their direction.

"Do you play music?", Feynman asked.

"Yes," Thomas replied.

"Good. I know a little joint further uptown a ways, in Harlem. They play lots of Caribbean there. They usually have a set of bongos for me. Maybe they can scare up some old bones and eye of newt for you."

Thomas was unsure how to deal with this less ominous "John Brown," who represented an apparently large, rather scary, and very secretive organization. He decided to play along.

"I'm sorry about the comment about eye of newt," Feynman said. I got a little carried away. You have passed some of my tests for compatibility with me, and with others that I work with. All that's left is your ability to convince me you have the mathematical skills and experience to do the work. More important, perhaps, is your ability to convince me that you know when to throw out everything you have learned about applied math, and your assigned task, when you discover – without any prompting from me, or anyone else in the unit – that your approach to solving a particular problem that I have assigned to you has gone to shit.

"Can you do that?" Feynman asked. "If not, we might as well split up now, and you won't hear from me again." "I can do that," Thomas said.

"Good. Let's go play some music."

====

Over the course of the evening in New York City, Thomas managed to at least make Feynman comfortable with being around him in a social situation. Feynman took Thomas into the subterranean depths of the City at 51st and Lexington so they could get on the right train for uptown on the West Side.

Feynman was suggesting they find a Hotel named the Braddock near the Apollo Theater in Harlem.

"The hotel has gotten a little seedy since the Harlem Renaissance of the 20s and 30s, but they have a good band playing in the basement," Feynman said.

"What kind of music?" Thomas said.

"Does it matter," Feynman asked. "They try and mix it up depending on the audience. They have a very good trumpet player, and they have some backup from New Orleans with brass."

"My instrument is a trumpet. Do you think they might let me play a few licks?" Thomas asked.

"We can only ask. If they like you the odds will be good. If they really like you, they might even loan you a trumpet."

====

Feynman and Thomas never made it to the Braddock. The day was Monday, August 2nd. On Sunday – the day before – a black GI had seen a white police officer who was trying to arrest a black woman for disorderly conduct. The GI sought to intervene. A scuffle ensued, and the policeman apparently shot the GI.

The Braddock was a block away from the Apollo Theater. There were bars, nightclubs, smaller theaters nearby along with several neighborhood hangouts. Columbia University a few blocks away on Morningside Heights was on summer break, but many summer school students had made their way down to the evening entertainment centers in Harlem. On hearing the gunshots a crowd of about 3,000 had gathered at the police station. A smaller group had gathered at the hospital where both the GI and the police officer were being treated.

Somehow a story got out that the black GI had died trying to run from the scene. That whole story later proved to be false, but the crowd had been growing angrier and angrier over several weeks with the increasing tension between the white majority that ran the city and owned most of the valuable property in Harlem. The black residents of Harlem who had to pay premium prices for lousy services from the City and local businesses in Harlem caused the crowd to swell further.

Among the black servicemen on leave who had come to Harlem to regain a sense of community to contrast with the anger over the hostile and persistent segregation in the barracks and chow hall, and the barely hidden racial hostility among white GIs was a further irritant.

A riot began that lasted for two days. It did not end until Mayor LaGuardia was able to show that the Black GI – a fellow named Robert Bandy – had not run from the scene, was alive and well, indeed he was barely wounded by the gun shot wound at the Braddock.

On Monday night, however, the word about Bandy's good health had not yet gotten to the rioters. By the time Feynman and Thomas got near the Braddock on Monday evening, LaGuardia had already mobilized several thousand officers and volunteers to keep the peace. Though able to keep the rioters more or less in check, by the end of it 6 were killed, 100s were wounded, and 600 were arrested. After rioters had decided to go home, property damage was mitigated, and hostile tensions lessened as Mayor LaGuardia ordered City Managers to get garbage, clean-up, and food distribution services moving quickly to restore order.

As Feynman and Thomas approached the corner of 126th and 8th Avenue, having gotten off the subway at 125th and 8th Avenue, several police cars approached the Braddock with sirens blaring. Feynman watched the excitement for awhile, but eventually realized that it would not due for him to be arrested and booked by the NYC police for any reason. Besides he assumed that the bar in the basement of the Braddock would probably be closed.

"Lets go on up to the Columbia campus. I have a friend up there in the math department. Maybe he is still at work, and we can still get a beer out of all this."

"I'm game." Thomas said, "but I sure would like to have had a chance to play the trumpet with a bunch of black guys from New Orleans."

Feynman winked. "Maybe next time."

=====

Pupin Hall is the primary center for physics research on the Columbia Campus. It is located on 120th Street that bounds the north side of the campus. Pupin Hall is about a 10 minute walk from the Apollo Theater In Harlem. The Columbia campus was calm relative to the sirens around the Apollo.

Feynman turned to Thomas. "I am going to check on some work we are having done here that will improve our ability to do our data computations. I cannot tell you any more than that, but I do want to introduce you to some of the people here. After I introduce you, I may have to ask you to leave us for awhile. If that happens, I will meet you at an Ethiopian Restaurant called "Massawa." It is on Amsterdam Avenue, a couple of blocks from here; …

1239 is the number I think. Here's some money to buy something to eat in case I get delayed."

"I have money, but do I want to spend it on Ethiopian food?" Thomas asked.

"Sorry. It is the only restaurant that I know this close to Pupin Hall, and the food is good."

"Where did you go to school?," Thomas asked. "Didn't you go here?"

"No. I once applied here, but Columbia had a limit on the number of Jews they let in, so I went to MIT."

Thomas was taken aback by this. He began to realize that he wanted to work with this man.

"If I get seriously delayed, don't worry about it. Go home and get some sleep. We will contact you."

Columbia had been exploring punch card methods for feeding large quantities of data into data processing equipment. That equipment was about the size of a stove used in a restaurant. Feynman thought it interesting that the punch cards were reminiscent of the punch cards used by Babbage and Lovelace to feed data to their steam powered Analytical Engine design 100 years ago.

In concept, the Analytical Engine design, had it been fully developed, would have been much more powerful – even if steam powered - than the Marchant manual calculation machines. The Marchants were heavily used by their human computers at Los Alamos, and the human computers themselves were growing increasingly frustrated with the complexities of manual computation, and the rising number errors. These errors needed to be caught, and fixed, and the data re-run. These deteriorating data handling processes threatened to bring the analysis critical to bomb design to a halt. The moving parts in the machines were wearing out to the point that they required constant attention by Feynman if they were to keep working at all. They were at the end of their usefulness and needed to be replaced very soon. Bethe assigned Feynman to recommend the kind of equipment needed to make a significant upgrade in the data capacity and the processing speed needed to more quickly produce accurate and relevant results from the configuration testing.

Feynman's visit to Columbia had a greater purpose than making introductions. He needed to check on the status of testing new ideas for speeding the processing needed for the huge quantities of data being generated by the work of the Theoretical Division.

This work was underway under the supervision of I. Rabi, and Harold Urey, both men well-known to Feynman from his earlier work with Rabi in Chicago. Based on what he would learn from Rabi and Urey, Feynman was authorized by Bethe to confirm a tentative purchase of several pieces of data processing machinery from the International Business Machines corporation. Once ordered, the machines would be installed and maintained by IBM on a priority basis in the computation lab at Los Alamos.

=====

While still working on the design of a nuclear fission system for an atomic bomb, Feynman took his Ph.D at Princeton. His thesis, published in 1942, and titled: *The Principle of Least Action in Quantum Mechanics*, explored a concept for simplifying ways to measure the yield of nuclear particle collisions at the subatomic level. In essence, Feynman believed that the tools for measuring the interactions of particles at the subatomic level were essential to the practical understanding of quantum mechanics. However, the tools currently in use were unnecessarily complicated, and, therefore, led to too many misleading results.

In advocating for a simpler way to measure the effect of colliding nuclear particles, Feynman was taking a position very much in opposition to the thinking of many scientists who were trying to understand the electro-magnetic forces discovered and defined by Michael Faraday, and explored further by James Clerk Maxwell, forces understood by Newton as "Action at a distance," and by Einstein as "spooky action at a distance." Faraday, Maxwell, and Einstein believed that these mysterious effects could only be understood as part of a surrounding "field." In his General Theory of Relativity, Einstein's concept of spacetime became one attempt to formalize his thinking about fields.

In his thesis, Feynman focused on the renormalization theory of quantum electrodynamics (QED). Feynman laid the groundwork for a consistent QED, free of meaningless infinities. He challenged all these ideas about fields, saying that they only added complication to the simple measurement of force when one nuclear particle collides with another nuclear particle.

"Summarizing:" Feynman wrote in the Conclusion to his thesis argument, "…a physical interpretation should be sought which does not refer to the behavior of the system at times very far distant from a present time of interest."

In other words - paraphrasing - focus on the question at hand; ignore all of the issues and history that brought us to this particular problem in the first place. This was Feynman's entire life and his entire attitude toward the solutions to problems that others had long since given up on. His Doctoral thesis merely reflected his most fundamental attitudes. When confronting a problem, ignore everything that has nothing to do with it, and be alert to all the possibilities for fixing, and re-purposing. Don't hesitate to try off the-wall fixes that might sound stupid at the time.

"Most important of all," Feynman instructed his staff, "try not to make a mess."

=====

A month after the incident at the Braddock Hotel in Harlem, and the visit to the computer testing lab at Columbia, Feynman called Sarah and new employee, Thomas, to a short meeting in one of the labs used by the Theoretical Division at Los Alamos.

"I assume you've met Thomas, Sarah?" Feynman asked.

"Yes. Of course." Sarah replied, then turned to Thomas. "I have been showing him around and introducing him to the human computers who are doing all our calculations while we wait for our new computers to show up. Maybe he has some thoughts he would like to share with us about what he has seen so far?"

"Yeah, thanks to you both, very much, for bringing me on to this project. I have been getting the whole story from the women who are working the Marchant machines about the kind of data you are processing, and the steps you have set up to handle the calculations. My compliments on your process arrangements for handling the data. It should all go quickly from paper onto the punch cards. If we can get those machines set up quickly we should be able to start processing real cycles of data within a couple of months."

Thomas continued talking. "I think I do have a few mathematical tricks I can contribute to the solution of the main problem you are dealing with now, the shape of the blast effects. You will have to merge the partial differential equations of Maxwell, with the statistical mechanics that can deal with the fuzzy positions of nuclear particles. Have I got that right, more-or-less?"

"My compliments to you Thomas. You have picked up a lot in a few days. I'm glad you are here," Sarah said.

"I'm also pleased, Thomas. I'm looking forward to seeing your work once you get your teeth into some of the real problems we're dealing with here. One more thing I want you to take a look at before you start locking yourself into approaches and methods that you think will work.

"Read my Ph.D thesis, and work through enough of the math to get an understanding of how I think we need to work out the identification and analysis of forces when particles collide in a condition of critical mass."

Sarah spoke. "Before we all get away, I got word from the hospital in Albuquerque that Kostanty has come out of his coma but is still suffering from loss of memory. Should we go down to see him, or should be wait until he is more able to talk to us?"

"We can go down to see him, but first I want you to take Thomas down to the explosives shack and introduce him to the sapper crew down there. I'd like to have his thoughts on what we need to do to speed up the turnaround on explosion data once we get our computers set up," Feynman said.

"I'm ready to go there now," Thomas said.

"Good." Sarah answered. I'll call up a jeep."

PANEL 9 - CHAOS

Rafaela, trained in medicine to become a Doctor, instead goes to Vietnam as an Army Nurse to find her cousin, an Army Green Beret. He is out of contact with his headquarters, he may be missing in action somewhere in the highlands near the DMZ.

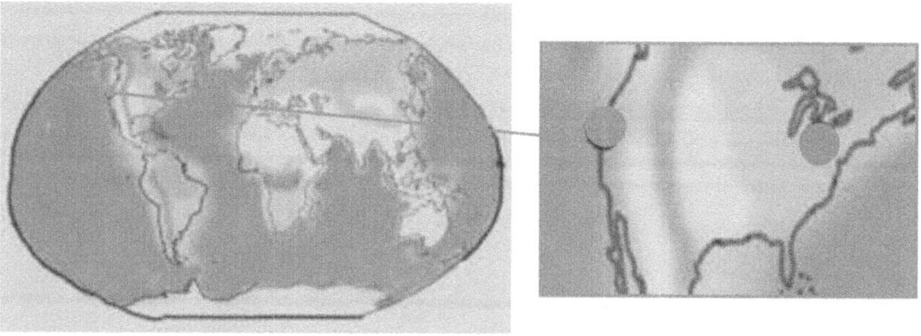

CLYDE, OHIO, SAN FRANCISCO, CALIFORNIA; VIETNAM DMZ;
SPRING, FALL 1967 CE

PROTAGONIST

Rafaela - An intelligent and talented woman in her mid-20s. She has just completed medical school but is taking a break before taking up the residency she needs to become a practicing Medical Doctor.

Matty. A friend of Rafaela since grade school.

Randall and Georgine Keagan Farm owners in rural Ohio. Georgina is a friend of Rafaela and Matty from high school.

Marina Cruz Another friend of the three women. She played basketball but not on the same team with Rafaela, Matty, and Gerogina until senior year.

Rafael - A close younger cousin of Rafaela. He is an infantryman with the US Army's Special Forces Units (aka the Green Berets) assigned to a small reconnaissance unit somewhere in the Vietnam/Cambodia/Laos highlands west of Hue.

Aleck Morris - A war correspondent for a small monthly journal, "The Prairie Observer," dedicated to reporting on employers who cheat their workers, and, lately on the Vietnam war.

Penny Blue - An editorial assistant on the staff of the "Prairie Observer." A friend and colleague of Aleck Morris.

Jack Raylance - pilot of a helicopter supplying food and ammo to the Combat base at Khe San.

Mike Fredrickson - A Green Beret who helps Aleck and Jack when their helicopter is shot down and crashes southeast of Khe San.

Fres Thao - A Hmong tribesman who works with Mike Fredrickson.

Chip Parker - A Marine Lieutenant from Georgia who knows something about the mysterious Rafael.

George Steiner - Aka Habit-forming, aka Hab, aka Tie-dye. A hippy picked up from a bulletin board message at the University of Chicago to help Rafe and Matty with gas money and driving on their trip west to San Francisco.

Colonel George "Butthead" Bukowski - Commanding officer of the Khe San Combat Base.

Captain David Margolis – Commanding officer of the medical unit on the Presidio.

Sergeant Collins – A Marine Squad leader on the Khe San combat base. He takes Aleck along on a "Lurp."

Hap. – a short stubby guy, a rifleman in Collins squad; Ralphie – a skinny guy with sideburns and, barely, a short scraggly beard. A rifleman on Collins squad; Sitting Bull – aka Bull. An American Indian sharp shooter and tracker on Collins' squad; Washington – a BAR man on Collins squad; Choi – a sharpshooter and sniper on Collins' squad; Pollock – a corporal, second in command to Collins on the squad; Easterman - radioman on Collins squad.

World War 2 ends in September, 1945. Many political and military leaders in the US and Europe did not rejoice. Instead they feared that Joseph Stalin's Russians, US Allies in WW2, had now gathered the ideas and technologies needed to finally build their own nuclear weapons. This meme had Russia, a mostly backward country now further depleted by war, trying to impose a Joe Stalin style communism on America and its Allies, now the most powerful nations on the planet. American military thinking pushed the meme that communism would come from the Russians with the support of the mainland Chinese under Mao Tse Tung. In a sort of confirmation of this, North Korean armies moved into South Korea on June 25, 1950.

At its peak, 327,000 American troops were involved in the Korean "Police Action." It ended, after much death and destruction on all sides, in a cease-fire. There were 33,686 Americans dead, but no resolution and no relief from the underlying issues and problems attendant on the "communist menace" meme. This meme has been at the heart of America's Joe McCarthy style anti-communist propaganda from the end of WWII well into the current century.

On August 2, 1964 The USS Maddox, a US Navy Destroyer on a CIA reconnaissance mission in the Gulf of Tonkin claimed that hostile North Vietnamese naval forces had approached them on the open seas. The Gulf of Tonkin is just offshore from the North Vietnamese capital of Hanoi. A few days later, official U.S. Navy reports claimed the Maddox had not only been approached by Hanoi's warships but had also been fired on.

U.S. President Lyndon B. Johnson had succeeded to the US Presidency following the assassination of U.S. President John Fitzgerald Kennedy (JFK) in November of 1963. Johnson was an otherwise brilliant domestic politician in the mold of U.S. President Franklin Delano Roosevelt but, unlike FDR, with little experience – outside the US Senate - in the management of large organizations, or the development of national and global policy. Worse, Johnson, formerly a school teacher in a small town in Texas, had very little experience of the people, the history, or the diverse cultures of southeast Asia.

Worst of all, he had almost no understanding of the post-war baby-boomer generation coming of age as voters around the time popular U.S. President John F. Kennedy was assassinated.

My magazine, "The Prairie Observer," had sent me into Vietnam a few months ago to get what you might call "local color" on the war. The magazine, and I do not see ourselves as "muck-rakers;" nor are we particularly biased politically, though we do try to see that deserving people got a fair deal from the people who earn profit off their work. The thing is that we love to write, and we love the process of journalism in finding people who know something about some difficult thing, and we know how to arrange interviews and ask the questions that might uncover a story of interest to a wider audience.

The pay was crap and the work intense and never-ending, but we couldn't think of anything better, or more fun, to do with our time in these days. Besides, the managing editor would usually let us follow our noses, and some of the female writers and staff were both talented writers and pretty hot.

Best of all, one of the prettiest of the editorial assistants, Penny Blue, was also a graduate in English from the University of Minnesota, and she seemed to enjoy my take on my reporting assignments. Penny's editorial re-writes of my work helps keep me clear, mostly, of problems with the editorial policies of the magazine's financiers and senior managers.

So far as we knew from the national news and wire-service chatter, the communists in the Vietnamese north, based in Hanoi, are trying to take over the legitimate government of the south based in Saigon. The most consistent policy line out of the White House was that the armies from the invading north were mostly rice farmers who would not be able to sustain their war missions in the face of a response by the U.S. military.

However, the battle for the Ia Drang Valley in November 1965 had caused some high-level people stateside to wake up to the difficulty of winning this war. My magazine decided they wanted to try to get ahead of it, so they could cash in with contacts and local knowledge as the war expanded. So, they sent me over to check it out, not so much because of my extraordinary talent, but because I was the only one with at least enough talent to keep my facts straight and write a simple English sentence. Besides I had something of a burning interest in the history being built in that remote and alien country.

Probably my assignment also happened because I am mostly free of domestic responsibilities and possible complications for the magazine.

We had been doing wire service "rip-and-read" news with press releases from Washington's official spokesmen on war stuff. But we wanted to become more relevant. Increasingly, we began to believe that we were missing the real news of the conflict. In early 1965, according to

Washington, America's military superiority and combat readiness "…will take care of these communist insurgents from the north within a few months." In late October 1965, the Ia Drang Valley brought four days of intense battle by 1,000 US Army regulars, draftees, and specialized combat units using new, helicopter-based air mobile tactics. They came up against about 2,500 ground-bound North Vietnamese Army troops. Even without the air-mobile assets, the NVA proved themselves to be highly skilled, dedicated, well-organized, and well-equipped.

After Ia Drang All the earlier happy talk in the US Media of an army of disorganized peasants and rice farmers began to shift. The NVA, though "defeated" – according to the U.S. news of the day - in the Ia Drang, had shown themselves to be a military force that might be more capable, and more dedicated than the North Koreans had been even when the North Koreans had the full support and a lot of manpower detailed from the Chinese army with them on the field of battle.

In speaking of the Chinese army, some of the US military planners who had planned the confrontation in the Ia Drang probably now wish they had paid more attention to the 1000 year history of Vietnam's many battles of independence from China. A few military planners at least realized, now, that the war in Vietnam was going to be long, tough, painful, expensive, and deadly.

There will be a lot of American casualties, a lot of politically unpleasant consequences, and a lot of bad press coverage. Whether found to be good or bad, I did intend to contribute my share of the coverage.

By the early spring of 1966, the calls for more troops and an expanded draft got louder. My draft number was high enough that I stood very little chance of getting called. By the time of my graduation from college in May, I had decided that I was not going to volunteer for military service. I was not much interested in spending a couple of years being ordered to shoot people or being ordered to get shot at.

Worst of all, for me, was the idea that I could get shot at, hit, and crippled for life as a result of bad decisions by the kind of people who would get

the US military, with all our bombs, bullets, napalm, and armor plate, into wars with people who live in mud huts and shacks, with roofs made of tree branches.

Even with these troubling thoughts I still felt pulled by the romance and adventure of war. As bad as we can imagine war to be, it still pulls on the imagination of younger guys like me who are filled more with the raging testosterone of youth, than with the reluctance and caution of maturity.

Following my editor's instructions on what I should do when I arrived in Saigon, I had spent a lot of time sitting on the rooftop of the Continental Hotel bar getting "acclimated." That meant sitting around listening to the same correspondents getting drunk every night and talking about their time "…in the shit." This got pretty tiresome. The more I listened to them, the more I became convinced that the loudest talkers had never been anywhere near the shit.

In early '67, Saigon was a long way from any combat "shit," though there were plenty of kids on scooters selling, or stealing, everything that a drunken, stupid, or horny GI might want to buy, or carry, or sleep with. The shit was in the streets of Saigon as far as I was concerned. Eventually, I decided to take my correspondent's credentials and try to find a way out into the boonies to check out some of the real action.

Later, I shared a few beers with one of the correspondents who talked and acted more-or-less straight up. He told me that if I really wanted to see some action I needed to hop a supply flight up to Da Nang or Hue near the DMZ, then hook a ride on a supply chopper to the combat base at Khe San. "Get to know some grunts and have them take you out on a 'Lurp' patrol. Get the commanding officer's permission to go along by taking his picture and promising to get it published back in his hometown paper."

After a pause, he then said: "You better make sure you really want to do this. You can get your ass blown off up there, and you won't even know what hit you." After another pause. he continued: "I mean it about losing your ass. You might wake up in a hospital with half your body gone, and the rest barely functional. Are you ready for that?"

"I can handle it," I said, "…but what the hell is a Lurp patrol?"

He looked askance for a moment then decided he would give me a straight answer, rather than some smart-ass response to what I learned, later, was a really stupid question. So far as most correspondents who had been in the shit were concerned, everybody in the World knows what a Lurp is, but I did not.

"Lurp is short for Long Range Reconnaissance Patrol. Lurps can be as large as a platoon of maybe 50 guys led by a Lieutenant or Captain, or as small as a squad of 7 guys led by a Sergeant, or even an enlisted man of lower rank. It depends on the mission; ...the distance, the objective, the need to bring along technical specialists and their equipment, the likelihood of enemy contact; ...stuff like that determines the manpower and firepower on a Lurp.

"Understand that a Lurp is the essence of infantry in war. If you want to understand war on the ground, you have to go on a Lurp; especially one that has contact with the enemy. Maybe you will be lucky and not get hurt."

"I tell you these things only because it looks like you actually might be able to talk your way into going along on a patrol; you might return with both your sanity, and your ass, intact; you won't have fucked up the mission; and you won't have gotten any GIs killed with some dumb-ass mistake."

======

It is about 450 crow-fly miles north from the main Tan Son Nhut airport in Saigon to the port and supply base at Da Nang, but our "Herc" supply ship had to fly a secure route because of increasing numbers of NVA ground units operating in South Vietnam. The several abrupt changes in direction to avoid ground fire, made the normal two-hour flight closer to three hours.

Once on the ground and shut down, the Herc pilot took me to the ops shack on the runway. From Da Nang it's a 100 mile flight to the combat base near the village of Khe San in the highlands close to the DMZ. The base is about 7 miles east of the Laotian border along the east-west highway named Route 9. In the shack I got lucky. The Herc pilot introduced me to the pilot of a supply chopper. After hearing my pitch, the chopper pilot looked at me with a quizzical look, but he said I could ride along to the combat base if I didn't mind sitting on ammo boxes and C rations.

The pilot called them "c-rats" which is what the grunts called them.

The combat base was originally established as a U.S. Army Special Forces airfield in August 1962. The base represented the initiation of then US President Kennedy's policy of military re-engagement with the North Vietnamese enemy in Indochina following the departure of the French in 1954. Kennedy's immediate purpose was to provide support for the US Ally, the government of South Vietnam in Saigon.

The government resided in the person of President Ngo Dinh Diem. Kennedy's military adviser at the time was General Maxwell Taylor. As the NVA began to infiltrate into South Vietnam, Taylor recommended that General Paul D. Harkins take overall command of US military operations.

President Diem was a member of the minority population of Catholics left over from the French occupation that ended with the battle of Dien Bien Phu in 1954. He had spent the summer of 1963 carrying out a program of brutal suppression of the country's majority Buddhists. There was so much hostility growing against Diem that he and his brother were both eventually assassinated by Diem's own military on November 2, 1963. Some believe that US officials met with unhappy officers in Diem's government in late October and had not discouraged the idea of an assassination of Diem. Even earlier, the self-immolation by Buddhist monk Thich Quang Duc at a busy Saigon intersection on June 11, 1963 – and the media's graphic video imagery that accompanied the global news coverage after the event - probably helped convince many that Diem had to be removed from office by whatever means necessary.

By the time of the Kennedy assassination, November 22, 1963, there were 16,000 "advisors" on the ground in Vietnam, including those at Khe San. Rumor has it that Kennedy never planned a full commitment of combat troops to Vietnam, and that he was planning to withdraw all troops as soon as it became politically feasible.

Now, the main US military guy for Vietnam is General Westmoreland, appointed by former Vice President Lyndon B. Johnson after Johnson took the President's office following the assassination of JFK. Whatever Kennedy had intended for America's military commitment to the war, there was very little talk of withdrawal in early 1964.

After the fierce 3 days of battle in the Ia Drang Valley in November 1965 Westmoreland had decided that bases in the area around Khe San would be a critical part of his plan to cut off the transport of war materials from the north to the south to supply the infiltrating NVA armies. He wanted a large contingent of Marines at the Khe San base to begin what he hoped to be the destruction of the Ho Chi Minh trail route for the supply of NVA forces in the south.

The head of the Marines in Vietnam at this writing is a General Lewis Walt. According to military scuttlebutt, Westmoreland and Walt had had a major head-butting contest in the Pentagon over Westmoreland's strategy of committing massive forces to the Ho Chi Minh trail near the DMZ. Walt was convinced that even if the Ho Chi Minh trail could be severed, or NVA movements of armaments and troops slowed down significantly, it would have very little effect on the gathering strength of the NVA already in the south.

After all, Walt may well have argued, the Ia Drang Valley is already 200 miles south of the DMZ, and the NVA was capable, even two years earlier, of fielding and supporting an army in the field that could stop a full-scale US Military operation using advanced tactics and equipment.

Whatever the truth of it, Westmoreland won the argument over strategy, and Khe San, supplied out of Da Nang and staffed by Marines, has become a major focus of the US military buildup.

As the Marines moved in, the original occupants of the combat base, the Army "advisors" known as "Green Berets", moved out of the combat base to Lang Vei about half the way to the Lao border. There, they continue to work with local Montagnard peasants (the "Nards") on training in US military tactics and reconnaissance methods. The Nards profess themselves to be politically aligned with the US interest in defeating the lowland north Vietnamese. They promise to help both US forces and the south Vietnamese army, the ARVN, in disrupting the movement of military supplies.

In their assurances of loyalty, though, the Nards may have understated their historic conflict with the lowland Vietnamese; …a conflict that has continued at least since the Vietnamese people established their independence from China more than a thousand years before.

Shortly after leaving Da Nang base, the pilot's voice came over my headset: "You better hunker down and hang on back there. The slopes are sending up some…" he paused, then, with a heavy French accent, "…fleurs du mal." I'm going to have to jerk the airplane around a little bit in order to stay out of trouble."

The pilot began to do some gut-wrenching dips, turns, and stalls with an airplane never designed for such things. It crossed my mind that the pilot might be secretly French, and an enthusiast for Baudelaire. Two loud bangs on the tail of the chopper followed by some ugly vibrations broke my escapist reverie on the ethnic origins and literary preferences of chopper pilots.

The rattling and vibration got more intense, and the chopper starting slewing from side to side. The pilot's urgent voice came over the headset. "They clipped us. My steering is a little fucked up, and I need to get her down on the deck." he said. "Check around your area and make sure you can see a way out of either side of the aircraft in case I fuck up, and we go in hot. If there are any loose boxes back there that could fly loose and hurt somebody, throw 'em out now. We are about 10 clicks out from the LZ at Khe San. I'll try to get somebody to help us out if the reception on the ground is too warm."

We skidded around the sky for a few more minutes. I am hanging desperately to my seat belt, and to the side of the chopper, praying we don't take any more hits. The pilot came on again. His voice sounded slow and gravelly this time. He is obviously wounded.

"Khe San is straight ahead. I can almost see the LZ (LZ means Landing Zone.). You need to look around the horizon now to check for landmarks so you can walk out of here. My arm is fucked up and I'm losing feeling in it. …and I am getting dizzy. I'll put her down in the next clearing I see while I can still control our descent."

We flew on for another 30 seconds, then, suddenly, the chopper flips 90 degrees to the horizon and starts to drop down, nose first. I brace, but I figure that this is pretty much the end of it. At the last second the chopper flattens out, the pilot kills the engine, and we hit the ground hard. The landing struts collapse. There are trees all around us, and the blades above us are still turning and slapping against some of the lower branches.

The pilot then slumps forward against his straps. He doesn't move.

I don't feel any broken bones, but I am sore as hell with bruises where I got banged into the chopper's metalwork and where one of the flying boxes has clipped me. I need to check on the pilot, so I unstrap and crawl forward over the disarray behind his seat. A sharp pain shoots through my left leg, and I let out a brief yell. I know my leg is not broken, but a corner of one of the boxes poked into it before flying out the door. I didn't notice it until I put weight on it.

In response to my yell, the pilot seemed to rouse himself enough to point toward the first aid kit on the bulkhead. I crawl over, gritting my teeth and favoring my sore leg, then tried to find his wound. His lower right arm was a bloody mess, he couldn't move it, and his flight jacket was shredded. His face looked pale. I have no experience with first aid beyond Boy Scouts and summer camp, but I figured I needed to stop his bleeding before I did anything else.

I grab a large roll of bandage out of the kit and wrap it several times around the pilot's upper arm, sort of like a tourniquet. By stopping the bleeding, I figured I would have time to try and help him get out of the chopper and onto the ground before he passed out again. Then I could try to find out how badly he was really wounded, and what more there was that I needed to do to help him.

Once I got him out and laying on some camo material on the ground, I checked his arm. The bleeding seemed to have slowed a lot, so I loosened the tourniquet and re-set it more carefully in the hopes that I could quickly figure out if the bleeding was arterial.

I need to decide what the hell I should do next. Jesus Christ. I am in the middle of the goddamn jungle in hostile territory trying to help a wounded soldier whose pure flying skill had just saved my life. I didn't agree to this when I signed on to come to Vietnam in the middle of a goddamn war. I asked the pilot how he was doing. He spoke not much above a whisper.

"Except for the pain in my arm, I'm OK for now, but there may be hostile NVA looking for us. There should be an M2 Carbine between the seats in the cockpit, and some of these ammo boxes should have ammo for it. Check around the cockpit for a 45 caliber sidearm; get the carbine, find and load the ammo, and I'll teach you something about fire-fights and perimeter defense. I have my own 45 caliber side-arm, but I only have a couple of spare ammo clips. Check the boxes for 45 cal clips and bring me all that you can find. Call me Jack, by the way."

"I'm Aleck Morris. …from upper Minnesota. Nice to meet you." I sound stupid. I even reach for his right hand to shake before checking myself. He smiled and reached up with his left. We shake.

"Wow", I thought, "…this guy has some balls." I crawl into the back of the Huey to see what kind of ammo and other weaponry I can find. We lost a lot as Jack went through the gyrations to dodge groundfire, but there were two boxes of carbine ammo, and a box of 45s. I also found the M2 and started moving the stuff over to where Jack could check it out and load the weapons while I continued to reconnoiter our situation and supplies.

Jack whispered hoarsely to me: "We could be here for a day or two. Check the chopper for all the C-rats you can find and bring that aid kit over. See if there is any water. If not, try to find something to catch rain water. I'm looking around for a place to hole up away from the chopper in case we start to get visitors."

I asked Jack if the radio in the chopper worked.

"Maybe, but it probably won't take our NVA friends very long to find us if we send a radio signal. We need to reserve that asset until we know for sure that it is our last resort if we are to keep from getting killed or taken captive."

What he said took my breath away. I had not even thought about being killed or captured. "Jesus Christ!" I said to myself again. This is real. This is as real as it gets.

Jack continued on: "I see what looks like a small rise about 100 paces from here. If there is some kind of a hollow behind it, we need to move our stuff and ourselves over there. Take the 45, make sure it has a full clip, then go check it out before you do anything else."

As I looked at the rise, and the hollow behind it, I started feeling hopeful that, from this hollow, we could at least hold off a few of the enemy if they decided to attack. Or, we could probably rest here long enough to regain the strength we need to be able to start moving toward our own forces around Khe San. With luck, the NVA won't find us, or they will decide to ignore us long enough for Jack and I to make our getaway.

When I got back Jack told me to start hauling the ammo and food over to our new hiding place.

"I'll stay by the chopper to cover you if any bad guys show up," Jack whispered. "Get moving."

"Don't lead the enemy to us by breaking a bunch of branches while you're moving this stuff. Step carefully and be thinking about how to either hide your tracks, or how to trick them into looking into places where we aren't. Once we move in over there, we won't be coming back to the chopper."

It will soon be dark, so it was time to stop talking and start moving our stuff.

=====

It took a couple of hours to move everything, to help Jack move to our hiding spot, and to do what I could in the darkening twilight to wipe out any sign of our movements. Jack is going in and out of consciousness, and I realize that I am dead tired as well.

Later, I wake up to total darkness, but some stars are shining through the trees. Jack is still either unconscious or sleeping. His chest is rising and falling normally. I have the feeling somebody is watching me. As I look around, I hear a soft whisper very close to me. I laid my rifle down next to me before falling asleep, but it was no longer there.

"I moved your rifle so you wouldn't accidently shoot me." The Whisperer said. "I'll give it back as soon as I am sure you won't do anything stupid like shouting, or thrashing around, or trying to shoot me. Do I have your word that you won't do anything stupid?"

"Who the fuck are you?" I whispered back.

"My name is Sargent Mike Fredrickson, RA 29244063. If it makes you feel better I'm from Brooklyn in New York City. Go Mets, fuck the Yankees. I'm in this shithole of a place as a Green Beret. We are kind of like bad ass Marines, only prettier. We speak the local language, and dress funny. I saw your bird come down earlier today, but I didn't want to let you know until I had a better feel for whether or not Charlie was in the immediate neighborhood, and whether or not you guys knew how to get your shit squared away."

"So, what did you find out?"

"So far, so good, but it pays to never let your guard down in this part of the world. You are lucky your pilot got his bird as far as he could south and east of Khe San. Most of the enemy is west of here, more toward the border with Laos. The guys who shot up your bird, though, probably saw you go down without a fire. They are probably headed this way to see if there are any goodies left that they can use."

"OK. I believe you. Can I have my rifle back? I'll try not to do anything stupid, but I have to warn you that I'm a journalist not a soldier." I whispered.

I still could not see the whisperer, but I now felt the returned rifle next to my hand.

"Yeah. I know that. You move like a pissed-off elephant at a rock concert. Worse, more like a bull in a china shop that has lost his balls to castration. If I am going to get you and your buddy back to a safer area, I am going to have to give you some OJT on stealthiness when moving and staying the fuck out of sight when not. How is the pilot doing?"

"I'm not sure. I am not a doctor or a medic either. He has lost some blood. I put on a tourniquet, but he seemed to be drifting in and out of consciousness before I fell asleep. He seems to be breathing OK now." "OK. Here's the deal. I work with one of the locals on these kinds of long range patrols. I dress like a local and speak the language pretty well. I don't speak at all when I am in hostile company. My partner is a Hmong tribesman named: Fres Thao. We don't carry noisy weapons on patrol, so I've sent him back to our base camp to see if he can round up some more help in case Charlie shows up here. He is a fast runner, and I expect him back with a report before dawn. His report will tell us what our real chances for survival are."

"How far away is this Help?" I asked.

"It is Probably 5 clicks to the Khe San combat base as the crow flies. Khe San is that way." He pointed a partially masked flashlight toward the ground in a direction that looked to me to be a little to the right of where I remembered the sun setting.

"Since the territory is likely hostile, that means it could take us two days. Think about how to get prepared for a long, fast hike with your wounded pilot," Mike said.

"I think you will be awake for awhile, so I am going to take a nap. If Fres comes back before I wake up I told him to wake me before announcing himself to you. I think your pilot will survive. He might need some morphine for pain, and some help from us, but he should be able to walk out of here. If you have any more questions keep them to yourself until I wake up."

Mike seemed to fade away like the Cheshire cat in "Alice in Wonderland." Other than his whisper, I still had no idea what Mike looked like; not his facial features, not the color of his hair, or how it was cut, not his clothes, or native dress, or whether he was wearing a hat, or had any weapons that make noise.

Even though it is the rainy season here, the sky was clear of clouds. After some time, I looked up to see if there were any constellations that I would recognize, and that might guide me in a safe direction if I and the pilot had to walk out on our own. Polaris was too far north to be seen through trees from this low angle. In the Southern sky, I was only familiar with the Belt of Orion, and the Southern Cross. Neither was visible through the trees. I thought of the Constellation of Cassiopeia, beautiful mother of the even more beautiful Andromeda, both punished by the Ocean God Poseidon for their arrogance. Cassiopeia and her throne upside down for half the night in the northern sky, across Polaris from the Big Dipper.

====

Rafaela is in Northern Ohio. The time is mid-May, 1967.

"I think I'll be headed to Vietnam in late summer," Rafaela said.

"Did you finish your military training?" asked Matty Franchetti, a friend from grade school. "Do you have time to earn some money first? Don't you still have a lot of bills to pay after medical school?"

The Cedar Point season had barely begun for the summer, but the day was hot. Both women had attended elementary school in a small farming community a little west of Castalia, Ohio. Both were sunning themselves on the beach. Cedar Point is an old, but still popular resort and amusement park from the 1920s at the end of a long sand spit connected with the south shore of Lake Erie east of Sandusky. The resort, known for its roller coaster ride is a 20 minute speed-boat ride across Sandusky Bay from the small industrial town in Northern Ohio of the same name.

Rafaela has recently completed medical school with an emphasis in emergency medicine, but she had not yet signed up for a residency. Since high school, Matty has worked as a secretary and executive staff assistant at the New Departure ball bearing manufacturing plant south of town. Unlike Rafaela, Matty had not gone to college after high school. She had taken a series of business courses at a local extension of Ohio State University and she had done well in them. Her quick wit, native intelligence, her ability to work hard and learn new business and factory floor procedures had earned her a lot of praise from a succession of managers at New Departure.

After high school, Rafaela and Matty had lost contact with each other while Rafaela had moved to the East side of San Francisco Bay to attend the University of California at Berkeley. When she was accepted into the medical school at Case Western Reserve University in Cleveland they reconnected.

Rafaela had a flair for travel and adventure. Matty was an admitted homebody. Despite these differences they found few others in northern Ohio, male or female, that offered the same kind of human and personal understanding, inter-personal challenge, playfulness, and awareness of the world around them.

They take considerable pleasure in each other's company.

They are often seen arriving together at such hot spots as exist around the Lake Erie islands and shoreline in northern Ohio in that hot summer. They are attractive, intelligent, and engaging women. Sometimes they leave these fun venues together, looking bored, or frustrated, or angry, sometimes they leave together in the company of men, or other couples; sometimes they leave separately, sometimes not.

Rafaela had been looking beyond Lake Erie toward the distant horizon. She lay back on the towel. She wanted to talk. "You know, Matt, there is a lot going on this summer out in the Bay Area around San Francisco. Some of it has to do with protests against Johnson's escalation of the war, but there is a lot of fun stuff too. I need some of that kind of mindless fun before I leave for Vietnam. Medical school was hard work, and my medical specialty in emergency medicine probably means I'll have a lifetime of it."

Rafaela twiddled the tie strings on her bikini bottom. She is an attractive woman with nice curves; tall with long legs. She had olive skin and dark kinky hair more typical of her southern Italian roots. She had more to say.

"I'm not even thinking about Vietnam, yet. I hear enough about it on the news every day. I'm still wondering if I should have volunteered to go. Then I think about my young cousin Rafael. He is in a special forces unit, and we worry because we don't hear from him, and the army won't give us any details; … only that he is on an extended confidential mission.

"A lot of people are coming into the City as war protesters, and there are already a lot of hippies there for the music, the drugs and the easy sex. The radical fashions and color and styling in clothes that you see around the Haight-Ashbury is exciting all by itself. "The more I think about it, the more

I think I want to go back there for at least a few weeks. I just want to see it, because it is so radical, offensive, and out-of-sight that I can't imagine that it will last very long.

"It wouldn't even have come into being at all, except that it is San Francisco; one of the most beautiful and fascinating cities on the planet. Want to go out there with me? You know, to kind of …check it out? I'm sure your boss will let you take a few weeks off."

Matty responded. "I could probably get the time off, Rafe, but whatever the U.S is doing in Vietnam has caused the demand for ball bearings to go up… fast. My boss has got me reconnecting with our suppliers every day to make sure our critical orders for materials stay on schedule. He also has me working almost full-time with the guys involved with the expansion of our production facilities," Matty replied. I'll think about what you are asking, though. It might be fun, and God knows I could use a break."

"Good." Rafaela rolled onto her belly, and untied the bikini top. "Put some suntan lotion on my back, Matt, please.

"You're such an animal," Matty said.

"After you've seen what I've seen in Med school, you'll realize that bodies are just blobs of flesh… nothing special, and nothing to get excited about until you and your team cut inside them trying to figure out how things go together and then work together with others – all of us covered in blood - so you can fix whatever is wrong before the patient dies."

"God, Rafe. You make it all sound so mechanical. Is it really as simple as reconnecting all the wires, and flipping a switch?"

"No. It's not, but maybe that's the real reason why I want to go back to the Bay. They tell me sex out there is getting exciting again. The idea of rubbing and touching naked bodies without having to worry about sterilizing the instruments appeals to me."

"OK. You can count me in. I'll work it out with the boss. How do you want to get there?"

======

A few weeks later, their outstanding local affairs and commitments resolved, Rafaela, Matty, and another friend from playing high school basketball, Marina Cruz, are at a dinner party on a friend's farm near the tiny farming community of Vickery, Ohio, 12 miles southwest of Sandusky. In the interest of saving money for more important things than transportation they had bought a used, fairly ratty 49 Ford sedan for their trip. They didn't improve the external looks of it much, but Rafaela, with Matty's help, did overhaul the engine, increased the displacement, added two four-barrel carburetors, a racing camshaft, and muffler bypasses on both tail pipes.

In this way, they hoped to occasionally make the trip more interesting, especially when dealing with guys or women they weren't interested in spending time with.

Both of the women had been handy with tools since an early age. Their families had expected them to help with all the farm work, including minor repairs and regular maintenance of the farm equipment. Even after being away from daily mechanical maintenance problems for nearly 10 years, they had to be considered among the best auto and equipment mechanics in the whole farming area between the south shore of Lake Erie, the city of Fremont, and the smaller communities of Clyde, and Castalia.

Marina Cruz is Hispanic. Her family had come up from Mexico to work the seasonal fruit crops further south. After a few years, her family were able to sign on with a local farming family. It had worked out so well that Marina's family was able to buy a small nearby farm of their own. They enrolled Marina in high school so she could play her favorite sport, basketball, and, perhaps, get good enough to play college ball with a scholarship somewhere.

Basketball scholarships were not much in the cards for any woman in the late 1950s, so Marina had enrolled in a local extension campus of Ohio State University while still in high school. She knew that the grades on her course work would have to be good enough to get into college with a scholarship of some kind or she would be back on the family farm slopping pigs and milking cows.

Marina had done well enough in her studies to graduate with a BA in teaching, and to have landed a job in nearby Fremont teaching bored high school juniors and seniors how to read and write. As an added benefit she had been asked to coach the school basketball team and had taken them to second place in a statewide competition in 1965.

Now, though, Marina was bored with her life. She hoped that by reconnecting with Rafe and Matt she could find a path toward something that would re-invigorate her interest in living a full life. The possibility of an expanding war in southeast Asia might offer some opportunities.

Other than Marina, Georgine Keegan had been one of the few people in high school that Rafaela and Matty would hang out with. She was intelligent and pretty. Even though she was shorter than the other players, she played a mean game of sandlot basketball. She set up the plays, her shots usually had some magic in them, and her aggressiveness and speed in re-capturing rebounds and loose balls was an example for all her teammates on the varsity team.

Randall Keegan, a local farmer, had admired Georgine from afar while she was in high school. He met her and proposed to her a few years later when he sensed that Georgine began to understand that life after high school could be lonely in the small towns typical of northern Ohio. Both seemed to be very happy in their marriage.

Randall and Georgine had inherited the farm from Randall's parents when they died in a car crash on the Ohio Turnpike a few years ago. As with

most farms in this part of Ohio, the Keegan's raised milk cows, hogs, chickens, and a few dogs and cats. They had about 300 acres of corn, wheat, and soybeans in cultivation, including pasture for the cows. They weren't probably going to get rich, but they work hard, and they make a comfortable living. Georgine was pregnant with their first child.

"Why in the world do you want to drive all the way across the country to such a place of sin and corruption like San Francisco, California," Georgine asked Rafaela. I thought it was bad enough that you were away from us for four years at the University there."

Rafaela wondered if Georgine was kidding with the "sin and corruption" remark. It seemed out of character. "It is not such a bad place, Georgine. I learned a lot at the University. I met some good people there, and I learned so many important things that I knew I was never going to know about if I had stayed here. Besides, you know I was getting crazy bored in high school. That's why I kept getting in trouble, and that's where I learned that the good news about high school detention rooms was that I could read any book I wanted there."

"Besides, San Francisco's mild weather, and rolling hills surrounded by water make it a beautiful place to be. People there are as fine as any people I have met anywhere. There are a lot of churches, and places of worship. Not all of them are Catholic, or Lutheran, or any denomination you might have ever heard about in Vickery, Ohio. Come out sometime and let me show you around."

Georgine suddenly looked down at her feet. Randall looked out the window as though checking the weather, though the warm, sunny, spring weather had not changed in a week. Matty's look told Rafaela that she had probably said something offensive. Even so, she had also wondered at the sudden sharpness of Georgina's remark.

Matty re-opened the conversation by asking Rafaela about her upcoming military service. Randall suddenly looked toward Rafaela, his steady gaze full of questions.

"What is that all about," he asked.

"I have been thinking about signing up, so I finally did a few months ago," she replied. "You know, my cousin Rafael joined up with the Green Berets after President Kennedy set up those special forces units when he was first elected. Rafael and I have always stayed in touch, and I have always been impressed with his commitment, and his belief that he is doing something important. Now that I am almost a medical doctor with some valuable skills, especially valuable in war time, I think it may be time I offered to do something important, too."

Randall was tight-lipped. "I don't know if what is going on in Vietnam can be considered a war, yet, though I guess LBJ wants to make it into in one as soon as he can. That business in the Ia Drang valley a year or so ago, certainly didn't solve anything. Johnson and Westmoreland threw everything they had at it, and they got their asses kicked by a bunch of rice farmers in pajamas and funny hats. Reminds me a whole lot of the "police action" in Korea. What a goddamn joke."

"Why hell, we are doing a worse job at defending 'American interests' than the French did at defending the right to eat croissants at breakfast in that shitty little country," Randall concluded, looked down at his farmer boots, and got up.

"Excuse me, I have to go out and slop the pigs."

"Wait a minute," Georgine said. "I'll come with you."

After they stepped out the door, Matty said: "Wow! Does any of that change your mind about signing up?"

"No." Rafaela said. "Maybe it is time for us to go."

As they walked toward their car, Randall called out to them: "Wait!"

Randall and Georgine both approached. Randall spoke: "Look, I'm sorry for being rude."

"I am too," Georgina said. You are both among our very best friends, and we have no right to judge what you are going to do. We want to always support you in whatever you do, even if we don't have a clear idea of why you are doing it."

Rafaela and Matty joined Randall and Georgine in a hug.

Randall took a step back and ran his hand through his hair a couple of times: "I don't know about this kind of war. I'm a farmer so I am always exempt from the draft and military service, but this kind of war, like Korea, seems different somehow from the war I grew up with in the early '40s. We are not sending our tanks and planes against our enemy's tanks and planes. We are sending all of the killing machinery and might of our military against a people who don't even have their own uniforms; …and for what reason is that again? Are we making the world safe for democracy by bombing mud huts? Is democracy and "We the People," no longer a consideration in these kinds of wars?"

Randall paused for a moment. "I fear that this war will end very badly because of that terrible premise. I hate to see my very best friends getting involved, but I also understand that because of the draft, many good young men are already involved." He paused and looked at Rafaela: "…and good young women like you are going even without a draft.

"Hopefully, these young people can balance the lunacy and the lobbyists for the armaments industries that I see driving all this war talk from our capital in Washington. You take care of yourself." "I'll take care of myself," Rafaela said.

"Me too," Matty said, "Though I'll be back in the neighborhood here after Rafe and I check out some of that sin and corruption you were talking about, Georgina."

Georgina burst out laughing. "I love you both," she said.

Randall didn't laugh, but he gave each a separate hug. "Take Care," he said. Marina had only met Randall today, and did not feel comfortable in group hugs with people she did not know. Randall bent over to look at her in the back seat. "Good luck to you too, Marina. Please come see us again," Randall said.

======

In early June, the sun was just coming up behind them as they drove toward the Rocky Mountains in the west. They were somewhere in the Utah desert west of Salt Lake City, driving fast across a long flat dry lake bed. Rafaela was driving. She had by-passed the mufflers, and her speedometer read a little over 100 miles per hour. They had picked up a hitchhiker at the University of Chicago to help with the driving. He had long scraggly hair, a tie-dyed t-shirt, and ratty old shorts and sandals. He also carried a large jar of multi-colored pills. He was asleep behind Rafaela. Matty was riding shotgun, also asleep.

Marina had appointments that she could not put off, so she did not join them for the trip west. She did say she might try to catch up with them in the Bay Area before Rafaela left the country.

Rafaela gloried in the freedom - and the sound, through her open window - of the screaming engine and open pipes echoing off the surrounding hills. Suddenly, she noticed the blip-blip-blip of a flashing red light in her rear-view mirror. She knew their souped-up Ford was fast, but it was about 50 miles to the Nevada border. She was briefly tempted to make a run for the border, but she probably couldn't outrun the patrol car before the Ford's souped-up engine blew up.

She pulled over, as did the police officer. The officer approached the car. He was calm but frowning. "Did you see the 25 miles per hour speed limit in the town you just blew through?"

"I'm sorry officer. I saw a couple of buildings, but I don't remember seeing a town. I know I am over the speed limit, but I'm a medical doctor, and I am trying to get to Fort Ord for some orientation to my duty station in Vietnam. I'm running late, and the road ahead looked clear and straight, and there were no cars coming." Whatever little fib it takes, Rafaela thought to herself.

"Do you have some papers that will confirm any of that?"

"Yes. I have a letter of congratulations from my Medical School. I also have a letter from the Army telling me where to go when I have my personal business wrapped up."

"Let me see 'em," the officer said. "By the way, who is your friend in the back?" He pointed at the scraggly form in the back seat, the bottle of multi-colored pills hugged tightly to his body, the bright pill colors clashing with the washed-out splashes of color in his tie-dye t-shirt.

"A hitchhiker we picked up in Chicago to help with driving and gas money. He told us the pills are diet supplements."

After looking at the papers for a moment the officer said: "OK. Look. I am going to do you a favor by letting you go with a warning, but please do me a favor by reconnecting your mufflers, and by trying to stay somewhere close to the speed limit while you are in Utah."

Rafaela felt a sudden flood of relief. She had gotten more stressed than she thought she would. "I will, officer. Thank you very much." The officer stood back from the car. Rafaela flipped the lever that reconnected the mufflers. She pulled back onto the main highway.

Matty lifted an eyelid. "Hey, Rafe. I'm really impressed."

"Cool," said the fellow in the back seat behind Rafaela; the one with the long scraggly hair and the tie-dye."

The remainder of the trip across the Rockies and the Sierra Nevada was routine, with the exception of a car wreck on one of the mountain passes west of Lake Tahoe. A car had gone through a railing and rolled down a hill. An ambulance and a state trooper vehicle were parked next to the break in the railing.

Matty was driving. "Pull up to that state trooper," Rafaela rolled her window down and spoke to the officer. "I'm just out of Medical school officer.
I might be able to help."

"Yeah, maybe so, but it looks pretty bad. Hold on for a second." The officer turned to a walkie-talkie; "I've got a young woman up here who says she is a med school graduate. Can she do anything useful down there?"

The connection has a lot of static, but a voice came through: "I don't think so. His neck is at a bad angle, and I can't get any vital signs. I don't think there is much to do but clean up and paper work."

"I'd like to go down anyway, officer."

The officer looked closely at her. He paused. "Knock yourself out," he finally said.

But there wasn't much anybody could do. Rafaela could see at a glance that nothing could be done to bring the twisted form back to life. She turned to the ambulance medical assistant standing next to the mangled '57 Chevy 2-door coupe. "You are right. There is nothing more I can do beyond what you have done, so I'll be on my way. I'll leave my contact information with the officer on the road in case there are any questions."

The ambulance crew thanked her for coming down the hill, and for being willing to stay in contact.

The three of them drove on.

=====

"Get up. We have to move… NOW!"

This was in the form of a sharp whispering directly into my ear. I rolled over to grab my helmet and rifle. Somebody grabbed my shirt and pulled me up. It was barely light enough to see, so I was having trouble keeping my feet under me and moving in the right direction. Mike was crouched and running ahead of me. "Where's Jack," I whispered.

"He's ahead of us. Don't worry, we have him", Mike said. "Keep quiet. Charlie's coming."

We kept crouching, ducking, and trying to move as fast as we could

through the mud and brush. The rising sun was behind us, so I knew we were running in the general direction of Khe San, but I had no idea how far we had to go.

After about a half hour of running, my feet begin to drag. Mike notices that I am about to trip over my own feet. He steps off the trail where the brush is partially cleared and pulls me into some denser underbrush.

"How you doin', cowboy?" Mike asks, his voice low.

I want to throw my weapon and extra clips away because they are heavy and are cutting into my skin, but I know I can't. "I'm out of breath and feeling totally beat up. How much farther?"

"My guide cut some brush for us, so as long as we can stay on this trail, we should be able to get to Khe San by nightfall. Otherwise we play boy scout for another night. "The only trouble is if Charlie finds this trail, he can move along it faster than we can."

"Do you know if Charlie is close to us?" I ask.

"No I don't, but if you stay quiet, I'll listen for noise that might tell us something."

I sat back to catch my breath. Suddenly it occurred to me why Mike is so hard to see in the dark, … he is black… black as the ace of spades. I had to ask: "I thought green berets were all white boys from the ivy league, not blackasses from the ghetto." I tried to keep my tone light, but I don't think Mike took it that way.

"Shut you mouth white boy. I'll talk to you in a minute about that."

I tried listening to the ambient sounds to see if I could detect any signs of trouble. But I couldn't. There were many birds, probably some monkeys chattering, and snakes slithering, but no human voice sounds that I could hear, nor were there the kinds of sounds that go with human activity; metallic clicks, bumps, scratching through cloth, branches breaking; nothing like that. Mike looked at me with his finger to his lips so I stayed quiet.

He put his hand on my rifle, then waved a finger at me to indicate I should stay calm. Suddenly, there was a disturbance in the brush a couple of yards away, and I almost crapped my pants. Then the leaves next to Mike opened slightly, and an Asiatic looking fellow stuck his head out. By now Mike had a death grip on my rifle. I couldn't have moved it if I had wanted to. Mike and the visitor talked quietly in the visitor's language for a few moments, then the visitor moved back into the brush, and disappeared.

"That was Fres," Mike said. "We almost have your pilot back to the Khe San base. He can move on his own, but he is pretty weak. Mostly we have had to carry him on a stretcher. You and I and Fres are on our own, but your pilot has a squad of Marines to take care of stretcher-bearing and perimeter defense."

"Where has Fres gone?" I asked. I wanted to know about our own chances for survival.

"Fres has been with the other group until he felt they were safe. Now, he is moving back along the trail to see if Charlie is anywhere around. If he finds Charlie, he may take him and his friends out. We won't even know about it. I think we are safe for now so lets move out as fast as we can. Stay low. How's your leg?"

"OK," I reply "but I don't think I can run very far, or very fast if we get into trouble."

"I'll keep an eye on you, but I may get ahead of you, and you won't know where I am. Try to keep calm because you need to keep moving. In general, I'll be moving toward the setting sun. As long as you move in that same direction, Fres or I will find you. If you lose sight of the sun, or start feeling disoriented and panicky, stop where you are, make yourself invisible, and wait. If it gets dark, get comfortable, and try to get some sleep."

"Jesus Christ," I thought to myself. What the hell was I thinking when I asked for this assignment. "Wait."

"Yeah?"

"What is the deal with you and the white boys in the Green Berets?"

"I do my job and the white boys don't give me any shit. If you can find one of them, why don't you do an interview. Maybe you'll get a good story out of it. C'mon. Enough talk. Let's move out."

"What do you mean if I can find one of them?"

"They are like me… survival out here means becoming a ghost… invisible to his fellow man. No niggers, no white boys, no chinks, no wetbacks among ghosts, my man. Now, let's move."

I did pretty well, but it did get dark, and I had lost track of Mike a couple of hours previous. I stopped in a place where I thought I could survive the night, covered myself with some branches and leaves. I fell asleep immediately.

The sky to the east was growing light when I felt a hand on my shoulder. I opened one eye and saw Mike. He motioned me to my feet, and then to follow him. After a few hours we slipped through the perimeter defenses, and into the compound set up around the Khe-San Combat base.

We are home, I said to myself …at least for the time being, and as much as a forward combat base can be thought of as a home.

====

Rafaela, Matty, and the tie-dyed passenger drove over the crest of the Coastal Range on Route 80 about 10 miles above Vallejo.

A local Latino rock station was on the radio. As they continued on south toward Rafaela's old stomping grounds around the City of Berkeley, the Golden Gate bridge came into view in the west.

"It's beautiful," Matty said.

Rafaela felt more at home here than any place else she had ever been on the planet. "Especially so in this cloudless day and beautiful morning sun," she said.

"This is really cool," said the tie-dyed one from the back.

"You know, Tie-Dye, we don't even know your name.".

"My friends call me Habit Forming, but you all can call me Hab."

Matty turned to ask: "What is your real name, Hab."

"Shit. I'm busted," said Hab. "My real name is George Steiner. What's yours?"

Matt turned to George and extended her hand. "I'm glad to meet you, George."

"Same here, " he said.

"My name is Matty, but you can call me Matt. Her name is Rafaela. You can call her Rafe."

"Are you guys new to the Bay Area?"

"I am," Matty said. "Rafe graduated from Berkeley."

"Cool," Said George. "I was born and raised in the City, over on Van Ness. My parents taught law at San Francisco State. I was in Chicago getting a

Master's Degree in Math, but mostly trying to beat the draft. That only works for so long. I finally got my notice. I have to report for Basic at Fort Ord in about a week."

"I know the feeling. I have to report to the Letterman Hospital on the Presidio in two weeks. Do you want us to drop you at your parent's house?" Rafe asked.

"No. My parents don't live here anymore. The gang I run with now are all over on North Beach. You can drop me off anyplace, but if you want to go over to North Beach I can probably get us something to eat, and some fun places to go this evening. Why are you reporting to Letterman?"

"I just finished med school, but I haven't done a residency yet, so I have worked out a deal where I go as a nurse, and basically pick my assignment. The Army said I should complete the residency, and get certified as an MD, then come back. The problem with that is that there are almost no female medical doctors anywhere in Vietnam – certainly not anywhere near any front lines, and I want to get to Vietnam… soon. And I want to get to the bush. I have a cousin there, and I have not heard from him for months. The Army only tells me that he is on an extended mission that is classified."

"Whoa, "George said. Where do you want to go?"

"The Highlands around the DMZ between North and South Vietnam. There is an evacuation hospital - near Phu Bai that is the first stop for those wounded in the highlands." The last I heard my cousin was assigned to a unit near the Laotian border… about a 100 miles west of Pleiku."

"Man. You got some set of balls, lady. I'm going to spend my time in this man's army trying to avoid doing anything dangerous, or even a little risky, or difficult in any way, and here you are jumping into the middle of the shit. Jesus Christ."

"Let me change the subject," Matty said. "What are some good places to go in North Beach? I've heard of City Lights Bookstore, Fisherman's wharf, Visuvio; are there other places?"

"There are many places. When you want to be entertained, no matter what kind of entertainment that is, North Beach is the place to go. You can find it all there, especially when you have a name like Rafaela. North Beach is basically an Italian fishing community."

"I'm going to drive over there," Matty said. "Can you show us some of the sights, Hab? Maybe introduce us to some people?" "Sure," he said.

Hab got the two women settled into a cheap room in a fisherman's retirement hotel near the Post Office. In the hallway on their floor, Rafe and Matt were greeted by a couple of old fishermen who were cooking a pot of fish stew. They welcomed the two women to San Francisco. They asked if they were hungry. The stew smelled delicious. "Starving" they said in unison.

After eating and helping the old men to clean up, they caught up with Hab, with his shirt off, down on docks in the Marina sunning himself. The three walked around for awhile then decided to take a cable car into downtown for some sight-seeing. Rafaela felt herself getting bored. Her mood did not improve on meeting some of Hab's friends in a lunch place down on Market Street. Some of them were stoned, and others had been in that condition not long before. Some looked as though they might have been stoned for the past several weeks.

They hung out for awhile at a bar and strip club that evening, but Rafaela said she was going to beg off on doing a late evening. She felt like she needed to catch up on some sleep.

Matty had been somewhat wide-eyed as she took in all the sights, sounds, and colorful inhabitants of the City at the beginning of this summer of love. At the sight of a very black man walking arm-in-arm with a very pregnant white woman; Matty's mouth fell open at the glow of happiness that seemed to surround them. Further along, at the sight of two Chinese men in vivid drag with high heels walking arm-in-arm, she almost swooned.

"There is nothing like this in Sandusky, Ohio," she said, sotto voce,

After awhile, Matty took note of Rafaela's mood. "I'm going to stick with Rafe," she told Hab. "Thank you for showing us around and introducing us to your friends."

Hab had been getting more stoned as the night progressed. He gave the two women a brief two-finger wave, and a lopsided smile. He turned back to his friends, who were equally out of it.

As they walked out the door Rafe noticed a poster for the Monterrey Pop Festival for the coming weekend in Monterrey, about 100 miles down the coast.

"As I get closer to leaving the states, I keep thinking more and more about my cousin Rafael. I just have a terrible feeling that he is in trouble over there, and I need to get to him. You understand, Matty, don't you?"

"Yes, of course, Rafe. Tell me what I need to do to help you. The problem is that I'm scheduled to go back to Ohio on Friday. If there is something you need me to do, better tell me now. On the other hand, if there isn't anything, then let's enjoy ourselves for the next few days in one of the great cities on the planet. Surely we can find something interesting."

Rafe pointed at the poster. "How about this Pop Festival in Monterrey? We will have to re-arrange our schedules, but it might be worth it." "I'm up for it. I think I can work it out with my boss," Matty said.

"I know a great Chinese place a few blocks from here. Let's go there and eat, and figure what to do. I know one thing I need to do: …After I visit the Presidio, I'd like to deliver our car to a friend on the faculty at Berkeley. I'm sure he can be persuaded to work with your schedule, and to drive you to the airport when it is actually time to leave."

Matty pulled her sun glasses down, crinkled her nose and stared at Rafe. "Is this a 'special friend,' Rafe?"

"He's pretty good in the sack if that's what you mean."

"Why don't we call Marina. We can see if she can fly out and go down to the Festival with us. " "Good idea," Rafaela said.

====

It's fall in the highlands, and it rains a lot.

I've taken a few days to recoup and try to catch up with my editors after the trip up from Saigon, and the helicopter crash back there in the boonies. Jack made it here, but they have medevacked him out to the mobile hospital at Pleiku. They tell me he is OK, but probably done with this war. I got to wish him well as they loaded him into the chopper. He told me not to take any helicopter rides just for the hell of it.

"You never know how those things are gonna go," he said with a big smile. We laughed and shook left hands.

====

I'm now sitting in a hootch that I share with a Lieutenant grunt named Parker from Georgia. His nickname is Chip. He is a combat commander responsible for Company C, 3rd Regiment, 26th Marines. He is telling me about a guy named Rafael. Rafael is born and raised in north central Ohio. His parents are part of a big Italian family in that area, most of whom work in some way with Great Lakes Shipping.

His old man was a longshoreman on the Great Lakes, and one of his uncles shipped out on a ship carrying taconite ore from mines in northern Minnesota to Sandusky, Lorain, Cleveland, and other ports on the south shore of Lake Erie. These ports are all connected, in turn, by rail with the steel mills in Pittsburgh and other places in the east. These are the places and homes of the people who build the tools, materials, and supplies demanded by the manufacturers who became the engines driving the Industrial Age.

=====

"I tell you this for background on this guy," the Lieutenant said. "Most of these Green Beret grunts are college kids who are looking for a little more action than you can get in the Peace Corps, but among the Green Berets, Rafael is truly a ghost among ghosts out here. Back in the world, though, he is a solid guy with solid roots in the Midwest. When you go out on this Lurp you've talked yourself into, I want you to keep an eye out for Rafael.

Now that I think about it, 'keep an eye out' is probably a misstatement. You won't see Rafael until he is on top of you. I could say 'keep an ear out, but, actually, that probably won't work either. You won't see him, or hear him, but you might smell him. He likes to wear a flower in whatever native hat he decides to wear."

"Well… you will also know if you feel his touch, but you probably won't like the feeling. If he doesn't know who you are, or why you are on his turf, he may put you in a death grip before he asks any questions." But I am sure you won't have any problems. Get some sleep."

Then he said: "One last thing. It is not like I know anything about his mission, but he saved my ass once, and I got to know him while we were trying get out from behind enemy lines. I don't want to lose track of him if there is anything I can do to avoid it," Parker said. "If he is in trouble and needs
my help, I want to make sure he gets whatever he needs."

Parker was right about my having talked myself into going on a Lurp. The combat base commander – a light Colonel known to the Marines under his command, but never to his face, as Butthead Bukowski - said I should get some sleep because he was not going to send me anywhere until he felt my chances of screwing up the mission was zero, and the chances of my surviving a crippling injury were at least better than 50 – 50… whatever that means.

"All I need is for some dog shit reporter from some chicken-shit anti-war rag in the Midwest to screw up my mission, and get killed, or captured, or physically fucked-up on my watch," is what he actually said.

If the "request" that I try to accommodate you hadn't come down from fairly high up, you'd be sittin' around here with your thumb up your ass until it is time for you to go home." He paused and gave me his best penetrating glare. "Got it?"

"Yes, sir, I got it. Thank you for setting me straight, sir," …I said. "Do you still want me to get your picture published in your paper back home?"

"Get the fuck out of here, Shithead," Colonel Butthead Bukowski said.

As it turned out, though, I sat around for several weeks with my thumb up my ass anyway. I also spent time hanging out with Lt. Parker, who I consider to be a pretty straight shooter. He gave me a lot of good background on what a combat base is and what a firebase does, and on the strategic importance of the Khe San combat base at this stage of the war.

There had been a lot of action at Khe San before I got here, but it was fairly quiet now that we were into the rainy season. The lull gave Parker and me a lot of time to talk in between my attempts to send some copy off to Penny over a flaky and irregular connection about the situation here. We also smoked the occasional doobie and toked on some opium whenever somebody brought some onto the base that they were also – rarely -willing to share.

He also gave me some basic training with the weaponry used on a Lurp. I spent a lot of time perfecting my aim with an M-16, blowing up bottles and stumps with hand grenades, and getting to know some of the grunts who actually do the ground-pounding on a Lurp.

====

It is now November 1967. Each day is generally dark and rainy. Chip and I have cracked some C-rats and are now chowing down. The USAF B-52 bomber air raids are continuing, weather permitting, along the Ho Chi Minh Trail, and we can still hear the bombs exploding. The B-52s can fly above the clouds, so they can still do some damage with radar guided bombs. Other aircraft, lighter bombers and our supply choppers can't fly in this weather. Supplies over Route 9 from Da Nang are also disrupted because of VC sappers working along the route.

This is all part of Westmoreland's strategy of disrupting the movement

of NVA troops and supplies from the north to reinforce their units in the south by cutting the Trail at the DMZ.

What a joke. Some of the scuttlebutt has it that the US has dropped as much ordnance on the trail as all of the ordinance dropped in Europe during World War 2. Once they figure out how easy it is for the NVA to anticipate those raids, and how easy to then move most of the hardware and personnel off the roads and into the woods, and to have the damage to the Trail repaired within a few hours, maybe they will stop disturbing our sleep with all the noise.

Our military planners in this war should have paid more attention to the details of the costs and benefits of the bombing campaigns against Germany in World War 2, and, specifically, one raid in particular, the October 14, 1943 raid on the extensive ball bearing works in Schweinfurt, Germany. For all the 300 B-17 bombers, crews, and bombs assembled for the raid by the Army's 8th Air Force, the production of ball bearings in Schweinfurt was back to pre-raid production levels in a matter of weeks.

Even the atomic bombs on Hiroshima and Nagasaki might have had the same effect on Japan's war-making machinery – virtually zero impact - as all the terribly destructive fire-bombing raids over Japanese cities that preceded them if it wasn't for the pure shock value of a single plane dropping a single bomb and causing so much total damage and fiery death, …Precisely none.

The thing that killed the Japanese war effort was the immediate surrender of Emperor Hirohito within a day or two after the US dropped its second, and last available mission-ready atomic bomb on the city of Nagasaki.

Whatever anybody else believes, so far as I am concerned the two atomic bombs had precisely no effect on the continuing desire of the Japanese military to fight to the death to, ostensibly, protect their homeland. In Germany, the war effort finally collapsed when Adolph Hitler committed suicide rather than surrender to the Russian armies closing in on his bunker deep underground in Berlin.

Now, two decades or so after Hiroshima, there is no "pure shock value" associated with the use of atomic weapons. The surviving eye witnesses to the blasts are now either dead or dying, and the narrative about the reasons for using the bombs then were mostly transactional. In other words, the rationale for using them at the time, in late July 1945 - according to then US President Harry S. Truman - was to reduce the loss of American lives that would result if the US invaded Japan with a ground invasion force.

There is no longer a moral or ethical dimension to the decision-making involved with strategic bombing; there is only the cost/benefit analysis; like being in a grocery store and deciding to buy chocolate or peanut butter. No wonder the grunts on the ground here are suffering such low morale; no wonder there is so much escape from reality. I am beginning to think that there is no reality here, only a faint disturbance in the global propaganda and mutual stroking that sustains these actions.

I'm sure that Westmoreland is considering the use of nuclear weapons on the Ho Chi Minh Trail, just to prove to his many critics - in the press, in the government, and within the military itself - that his strategy is correct in dealing with his military problem; a problem so severe, Westmorland believed, that the use of nuclear weapons is justified. If he did get approval from Washington to use them, I just hoped he would do us the courtesy of giving us news guys and the grunts around the Trail enough time to get out of the way.

====

Colonel Butthead finally gave me permission to go out with a Lurp. The mission will involve a squad of Marines from the Combat Base doing a reconnaissance along a trail to a Green beret outpost at Lang Vei. Lang Vei is very close to the Lao border. We leave at 04:00 tomorrow morning. Butthead told me I would have to gain the trust of each of the members of the squad before I could go with them.

"If any squad member doesn't want you to go, then you don't go. Got it?"

"Yes, sir. Got it, sir," I said. "Where do I find the squad?"

"They are gathering their gear and supplies over next to the Ops shack. You better get over there."

He started away then paused and turned back to me.

"One last thing, …I have been impressed with the work you have been doing with Lt. Parker. He is also impressed with the improvement in your weapon handling, and in your physical ability to handle the sheer hard work involved on a Lurp. He has been your strongest advocate," Butthead said, "…and he has agreed to go along to look out for you. You better get a move on. Find Parker and take him with you to help break the ice with the squad."

Parker caught up to me as I made my way to the Ops Shack. He has removed his insignia of rank. The squad members are in battle fatigues, and are already packing up their gear, getting ready for the early morning start. Parker introduces me to the squad leader for this mission, Sergeant Jerry Collins. Collins is a big guy… at least 6 – 3 or 4. He shakes my hand but doesn't say anything. He points to the other members of the squad, then gives a nickname as he points to each one. A short stubby guy is "Hap." A skinny guy with sideburns and barely a beard is "Ralphie." Another big guy, more heavy than tall, is an American Indian named "Sitting Bull; …sometimes just Bull."

After the introductions, Collins turns to Parker, and asks if he wants to say anything about me.

Parker tells them he has been working with me and has been training me on weapons and fitness. "I'll be going out with you guys to help Aleck get his story about what we do, and to know how to keep out of the way while you do it. He's a good guy, and he has been working hard to be able to do this with you. I do outrank Collins but he is your commander on this mission. Don't forget it. If I am forced to remind any of you to look to Collins for leadership if things get tricky, I'll remind you, once again, that Collins is the boss, and I will remind you in ways that you won't forget."

"Give Morris something useful to do on the Lurp. Otherwise, he will start thinking this is just a long hike like the ones he used to go on holding hands with his sweetheart back in Minnesota. Give him some gear to haul and be responsible for, and make sure that he gets a weapon with a couple of clips to use if things get bad in the shit. I have been training with him. He knows how to handle your weapons, he understands basic tactics, and he is a better shot than a lot of the other combat grunts I know on this base."

"Any questions of me or Morris?"

One of the black guys, "Mary Jane," asks: "What the fuck you doin' here, man.? You don't even have to be here in this shit hole, and you all are going out on a combat mission? You fuckin' nuts, man. We're gonna get our assess shot at, and some swingin' dick's gonna get hurt; …guaranteed. "

I didn't have any quick answers for Mary Jane, but I knew I had to say something. "Look. I know you see me as probably more of a liability than an asset to this mission. All I can tell you is that I have been trying to get myself into shape, and to learn how to handle weapons if we get into the shit.

"I'm trying to do the best I can to understand what you guys do, and what you have to deal with out here so I can tell our readers back in the world what this war is actually all about. Going on this patrol is probably the most important thing I will do while I am in Vietnam, and I promise I will do everything I can to not let you guys down."

Another guy with Asian features – name of Choi, a weapons guy - asks what I will do if Parker and Collins get into a pissing contest during a firefight. "Who you gonna go with?"

"Collins." I answered with no hesitation. Another guy, Washington, the sharp shooter, butted in: "Is Parker your baby-sitter?"

"I don't have a baby-sitter. Collins is squad leader. Collins is the boss."

From the back: "Good answer." A guy named Pollock smiles my way and says: "Not common over here in the shit."

Collins speaks up. “Let’s cut the bull and get you some gear. Pollock. You help him out.”

“Got it, Sarge,” Pollock replied.

“Sounds good to me,” I said.

=====

“Monterrey is a beautiful city. The views from above the beach at sunset will stay with you forever”, Rafaela said.

Marina had come out from her small community in Northern Ohio. She had never been in San Francisco and had not yet been downtown on this trip. Rafaela and Matty had picked her up at the Oakland airport, and they left immediately for Monterrey.

But Marina knew the area, particularly the farming areas south of the Bay because she had grown up working in them when her parents and family came up from Mexico to work the fields during the growing season.

The hard work and abuse by field owners against all the Mexican families had not left Marina with good feelings about the people of these farming areas. Even so, her family was very loving despite the hardship and abuse, and they were able to protect Marina from the worst of it. More importantly, as Marina grew steadily into the beautiful woman that she had become she began to attract a little too much unwelcome attention from some of the field owner’s male children.

Marina protested her mother’s protectiveness, and her father’s private threats against anybody who got too familiar with her. Marina had always had a sense of pride about her body, and she took care to run several miles each day and do other exercises and workouts with weights. She felt she could take care of herself, and loudly and repeatedly said so to her parents. As a family, however, they all knew – even Marina – that it was time to leave California and make a new home for themselves in a different area of the country; one where their hard work and strong family values would be more valued than exploited.

Marina would be forever grateful to the family in northern Ohio who had taken them in and had made it possible for them to buy their own small patch of farmland. She felt very grateful, as well, to Rafe and Matt. Marina’s athletic skills would have languished and faded if she had not had them constantly sharpened by competition with these two talented women from a nearby school that regularly played against her school.

Now the three grown women were together in a beautiful place with beautiful music, and they intended to take full advantage of it.

Marina's flight had got into Oakland by mid-afternoon on Thursday. The three decided that they would leave for Monterrey immediately in the hopes of learning how to find their way around and finding a decent place to stay close to the Festival venue. Despite their best efforts, they did not find a decent hotel. Rather than sleep in some flea-bag a ways out of town, they decided to find some space on the beach to camp out.

Even the beach was crowded with transient Festival-goers who could not afford to, or did not want to buy hotel space, but who were in the mood for a big party. For the three attractive women from northern Ohio, the night was full. They got up late the next day, and spent some time finding something to eat and deciding what Festival acts they were most interested in.

Over a cheesy croissant, Matt suggested that "Big Brother and the Holding Company" were featuring their incredible female singer Janis Joplin, and they would be on-stage early Saturday morning. "The George known as Hab told me that she and the band have been playing the clubs in the Haight, and her singing will literally '…blow your socks off…' is how I think he put it. I think if we do nothing else here," Matt said," we need to hear Janis Joplin sing."

"OK. I'm in," said Rafe. "Me too," said Marina."

=====

Back in San Francisco at the Presidio on Monday, Rafe drove to her appointment with a Captain Dave Margolis, head of the medical unit. He grinned and shook her hand. He invited her to sit down. "Welcome, Doctor.,"

"I'm not quite a full doctor, yet, though I hope to be one as soon as I complete my obligations here, and get into a residency," Rafe said.

"Well," said Margolis, "I hope we can help make that possible. Your papers are squared away, and I've got clearance to put you on a bus to Fort Ord tomorrow. Are you ready to travel?"

"Yes sir, I am."

Ask any soldier you see where it is. You will do about 90 days of training on how to look and act like a military Lieutenant who is also a military nurse. Once that is done, as I understand it, you'll be flown out to Saigon, then dispatched out to the 22nd Surgical hospital at Phu Bai up by the DMZ. The commanding officer at Phu Bai is Captain Tom Burgess from Tuscaloosa. He graduated from Old Miss. He is a good guy, but he talks funny and can be rough around the edges. He has a lot of experience with combat trauma, though. He has been briefed about you and is looking forward to the medical help you can provide. The nurses up there tell me he knows his stuff, but he can be a real shithead about procedure if you get on his bad side. Unless you have any more questions of me, Good luck and God speed."

====

Collins was pissed. The radio was fucked up, and the radio guy – Easterman - couldn't call in an air strike on an NVA position set up below them across Highway 9, about a mile west.

The NVA was too far away for our sharp shooter, so Collins determined to get closer. I wasn't too happy at that idea. I guess my recent experience in the deep bush in hostile territory was still lingering with me, but I really couldn't say anything about that, and I really couldn't say that I didn't want to go. I started picking up my gear to move out.

Collins spoke up. "Things could get hot, Lieutenant. You and Morris may want to stay up here while we take care of business down below. We'll get back to you as soon as we can."

Parker looked in my direction, but I kept my face expressionless. I continued gathering my gear, but a million thoughts are running through my mind. Then Collins looks at me, and I blurted out: "I need to come with you, but I don't want to screw up tactics, or get caught in any cross-fire. Is there a place where I can see what's going on, and you guys can find me after it is over?"

Collins looked down for a moment, then to the west. "You see that rock outcrop ahead of us about a half click from here? You have to stay behind us, but you can follow us so long as you can keep us in sight. If you lose us, head for the outcrop. We will get there when we get there. If we don't get back to you, figure you are probably on your own. Plan on staying the night, then make your way to Lang Vei at first light tomorrow morning. There is a firebase there with a bunch of jungle bunny Green Berets. If you can make contact with them, they will get you back to Khe San.

"The main thing to remember is that all of Route 9 is contested territory. We might control a piece of it one day; the slopes will take control of the same piece the next day. Typically, what either of us does when we first retake control of a piece of the road is hide and set up an ambush. Before you show yourself anywhere on the road, stop, look, and listen. Don't make a move until you know for sure who is in control.

"All those B-52s that have been bombing the shit out of the trees and dirt along the Trail haven't really done any more than stir up a big hornet's nest. NVA troop strength is building in this whole area every day. So, watch your ass, stay in the woods and stay off anything that looks like a regular trail.

"We gotta go," Collins barked.

Mary Jane, the black guy sat nearby. He had an ear cocked in our direction. "Uncle Ho has figured that Uncle Sammie thinks the war can be won by bombing the Trail. The Ho Ho man wants to send enough troops and weaponry into this area to keep Johnson and Westmoreland thinking that the bombing is the only way to stop the NVA providing support for their NVA buddies in the south.

"If the NVA collapses in the south, why, sheeeit, man, we can be home by Christmas." Mary Jane broke out into a raucous laugh, then he stopped. "What a fuckin' joke."

Mary Jane looks like he has more to say, but Collins cuts him off. "Let's go."

Parker grabs his gear. We wait for a few minutes before following the squad down the hill toward the NVA position. It became clear right away that we could neither keep up, nor keep the squad in sight. They had disappeared almost immediately. Collins' comments about the NVA buildup in the area also began to work on our minds.

Parker spoke first. "We need to clear our heads about what, exactly, we are doing. We can't be anywhere near that squad if they get into a firefight. If Charlie finds us, we can't survive on our own away from the squad anyway, even with the weapons we have, even if Charlie does something stupid... which he is not probably going to do.

"We need to move now to the outcrop and see what happens from there. Maybe we will get lucky and have a good view of the NVA position so you can get your story."

=====

I was familiar with this kind of bush-whacking because of my escape after the helicopter crash on the way up to Khe San with Jack. The number one rule: stay away from anything that looks like a trail, because it is probably booby-trapped. Booby-trapped or not, Charlie will recognize your footprints, and will come after you if you give him a trail of shoe prints. Remember that Charlie is probably wearing old worn out shoes that may have been repaired by strapping pieces of old tires together. He might track down his own mother if it meant he could get a new pair of shoes.

We are now a couple of hundred yards from the outcrop. Parker is leading. He signals for a break, and we sit down. The bush is fairly dense here, so we can't see much.

Suddenly, an oriental face appears in front of me with his hands up, and his palms out toward me. His face was no more than two feet from mine. Immediately, I recognize Fres, the Hmong tracker with Mike Fredrickson, the Green Beret who helped get me the rest of the way to Khe San after the crash. Fres had his finger to his lips, so I reached over to touch Parker, who had his back to me. He had not yet seen Fres. I didn't want Parker to suddenly turn around and see an enemy.

Parker was astonished, but Fres and I both had our fingers to our lips, and our other hand up with the palm toward him. He looked at me with a quizzical look, but he did not pull his service revolver. I deferred to Fres, who then indicated that we should stay put and wait. Fres then turned around and disappeared into the brush.

About 10 minutes later, the bush parted once again, and Mike Fredrickson appeared, and whispered. "Hey, Guys. What's up?"

"Hey, Mike. Its good to see you again… I think. This is my pal Lieutenant Chip Parker."

"Hey Chip, hows it hangin'?"

Mike didn't wait for an answer from Chip.

"Here's the deal. You guys have busted into one of our ops. We are trying to take out that NVA installation down there just off the highway. They have been doing a lot of damage since they set up a few days ago. We were ready to do it this morning until we saw you guys come on the scene. Rafael is trying to get to your guys before they all get wiped out. Charlie has probably been watching since you left Khe San; they probably have the ambush set. They might be watching you – actually the four of us - now." Parker spoke. "Did you say Rafael?

"Yes, I did. He is one of our best. The only reason I use his name is because you know him. And he knows you."

Fres punched Mike on the shoulder and pointed toward a truck coming up on the highway. It appeared to be Russian, which means it is NVA.

"We've got to get down to your team. C'mon. Try and keep up. Fres will help you."

Mike disappeared into the brush. As the truck got closer to us, gunfire broke out from what I knew to be Collins position. Parker and I followed Pres down the hillside, trying our best not to trip on the underbrush. I caught a glimpse of the NVA position on the highway, and noted that they were looking toward Collins, probably trying to find where the gunfire was coming from. One of them pointed a rifle at Collins position, then dropped as though shot.

An NVA machine gun opened up and began tearing up the hillside around Collins. Another rifle opened up near the Collins position, and took out the NVA machine gunner. He was quickly replaced, and the new gunner continued to fire.

"That's Rafe," Mike said. "He's working alone. I have to get down there to help him out. Fres, see if you can get Parker and Aleck to Collins position. If Collins has a radio, try to get through to our guys, and get them out here ASAP, and ready to do some evacuations, and some hunting. The weather is shitty, but it might be breaking up. See if you can raise a chopper from Khe San to help us out. We may have casualties."

Parker spoke up. "Wait. I owe Rafe. I promised myself if he ever got in trouble, I'd give him all the help I could. I'm coming with you. If you need me to lay in some covering fire, while you guys do your magic away from me, I'll do it. Morris is good in the bush, and he knows the weaponry. He can help Fres take care of whatever Collins is dealing with.

I took note of the truck and was glad to see that it was backing up, trying to get away from the firefight. That's good news… it says they are more transport than combat. It takes Fres and I about 15 minutes of bush-whacking to get down to Collins. Collins was out with a head wound. I looked for Pollock.

"We've taken some hits," Pollock said. "Mary Jane and Hap are gone. Collins is functional, but groggy. The only good news I have for you is the radio. I think Easterman has almost got it going again. We need some help, and the only way we can get it is the radio. If either you or your slant-eyed pal can get that radio to at least get some kind of a signal back to base, I'll put all you greenies in for some kind of a medal, or at least a beer back in your hootch in Lang Vei. Even you, Morris, …and your pal, too."

I had gotten to know Easterman on the way here. Whatever had happened to the radio was probably caused by rainwater getting into the works, rather than any operator-error problem caused by Easterman. We found him busy drying parts and connections. Fres helped with this. In his time working with the Green Berets, Fres had made himself useful by working with the electronics. His English was still pretty rocky, but he was a fast learner, and he jumped right in with Easterman to dry parts.

Pretty soon they were able to get the radio to light up, but there was no real way to test the sending signal except to start sending a signal for help on the channels for Khe San and Lang Vei, and keep repeating it

Pollock came over to us. "I'm seeing some action from the NVA across the road. I think they are calling some heavier firepower into our position. That is bad news for us. Unless the shooters are able to take them out, our only hope is that your signal is getting out, and the white hat guys can get to us most ricky tick. Is the radio working?"

"Yes. Looks good so far," Easterman said. "We are getting a signal out."

A minute later, an artillery shell came over and blew up some trees about 100 yards above us. Dirt, branches, and rocks started falling on us. Easterman spread his arms over the radio to protect it. The NVA were bracketing our position with artillery shells. A couple more shots and they would have us. Washington and Choi, the BAR man and the sharp shooter were scrambling to get to firing positions to the left and right away from the squad and the radio. Washington began firing the BAR in short bursts, while Choi concentrated on repeated aiming and firing.

I looked down and suddenly saw that Rafael and Parker were scrambling to cross the road, covering each other while one or the other tried to get close enough to use grenades and hand-to-hand knife-work on whoever remained at Charlie's position.

Somebody in the NVA position was calling in corrections to their artillery. The sharp-shooters knew they would have to find him and take him out before he was able to guide the artillery to Collins and the radio.

=====

Yesterday, I woke up groggy and bandaged up. A nurse came running up to my bed with some kind of a needle.

At least I think it was yesterday. It might have been last week. The nurse jabbed me with it, and I fell back to sleep. I feel a little more alert now but I have no idea where I am, or what happened to me. Last I remember, an artillery shell was heading my way during a firefight somewhere out on Route 9. I am not in a regular hospital though. I am in some kind of a tent with several beds. Some of the other beds in my area are occupied by others with war wounds. I guess that this is a field hospital. I hear gunfire in the far distance, so I suppose we are near some action, but I don't know where. My head and left leg are bandaged. I can feel pain in several places on my body. The pain in my head is intense. My left eye is covered with a bandage. I can see and hear, and my thinking is clear enough that I am less worried about the severity of the injuries.

Most important there are no bandages and no pain down near my private parts. Thank god. I have dodged "The Wound to end all wounds" … the diagnosis no GI ever wants to hear. Scuttlebutt has it that a GI will give up any other body part to a combat war wound, but not those little babies between his legs.

A female lieutenant with a stethoscope approached my bed. "Hey soldier. Welcome. If you are really awake, I'll bet you have a lot of questions. I'm going to take your temperature and check out some vital signs while we talk."

"OK. First, I'm not a soldier. I'm a correspondent with "The Prairie Observer" a little review of current events and news out of Minneapolis. My first question is who are you?"

"I'm Rafaela. I'm new on the job here, but I do have a medical degree, and I do plan to do my residency for a real MD when I get out. How in the world did you get yourself attached to a combat mission as a correspondent?"

"I had to work for it, but, like I told the squad I went out with, going out with them on a Lurp is the most important thing I will do as a correspondent here. And I promised them that I would write it for the folks back home in a way that would help them understand what this war is really all about. I can't believe my good luck in still having all my parts, and most of my wits after getting wounded like this."

"Yes. You are very lucky."

Rafaela has beautiful eyes, dark skin, and beautiful dark, curly hair. More important than her looks, she has a way of looking at me that is totally engaging. She is going to be a hell of a doctor, but I discovered that I was getting a hard-on. I was torn between hoping she was not embarrassed by it, on the one hand, and, on the other hand, wishing I was mobile enough and knew enough about her to make some moves on her. "Do you know anything about the other guys on my squad?"

"Yes I do," she said. "How much detail do you want?"

"I recall that two were already dead by the time I got to their position. The squad leader, a Sergeant named Collins, had been hit, and was pretty much out of it. Pollock and the rest of the squad was basically functional. Two sharpshooters were moving into better firing positions, but I don't recall them shooting at anybody yet. I was working with the radio operator, Easterman, trying to get a signal back to base for some help. There was a Hmong by the name of Fres Thao also working on the radio.

"Fres was working with a Green Beret named Fredrickson who helped me get out of a bad situation when I first came up here from Saigon. There was a lieutenant named Parker who had been helping me out, and was with me on the Lurp, but he went to help another Green Beret named Rafael who he owed a favor."

Rafaela eyes grew wide at the mention of the name Rafael. She became short of breath, and her talk was suddenly rushed. "Rafael is my cousin. We are both from Ohio. I am actually here because I am trying to find him. He has been out of touch with any of my family, or any of his friends for several months. All we can get out of the military is that he is on an extended confidential mission. You might be the first person I've met here who might have actually laid eyes on him."

"I'm sorry I don't know more." Actually, I did know more, but I am reluctant to tell this particular nurse anything about it. "I'm sorry, I don't remember anything after that except the noise of an artillery shell that was headed my way."

The nurse Rafaela had finally managed to relax as I related all I knew to her. "That's OK. You have given me some hope that he might still be alive. I have to caution you, though, about saying anything about Green Berets, or their operations in this area. I don't think their presence here is a secret, but the things they do on some of their missions probably are. Just be discreet. I'd like to talk to you more, but I've got to look in on some other patients. I'll be back."

As she began to leave, Rafaela turned back to me: "By the way, you are in a temporary field hospital at Khe San. The weather and enemy artillery and other actions are preventing evacuations for now. We do hope to get you and others out of here to Da Nang in a day or two. Your friend Easterman, the radio operator, is two beds down. He hasn't returned to consciousness yet, but he will soon. If you both like cribbage I will see if I can rustle up a board."

"Just one more question. Do you know how I got here from the firefight?" I asked.

"No I really don't. All I know for sure is that one of the Air Mobile helicopters came in hot with some damage from ground fire five days ago. You and Easterman were on the floor of it, unconscious. Between the weather and hostile fire going out and coming back, according to the pilot and gunner, you were all lucky to get here at all." "How about the others?" I asked.

"I don't know. Sorry." Rafaela replied. "The pilot went out again, but he didn't come back this time."

"What happened to Fres? Do you know?" I asked.

"There was no mention of any natives, Montagnard, Hmong, or otherwise, being involved in the action you were involved in." Rafaela said.

I thought for a moment as Rafaela turned to go again. "Wait." She paused. Then I said "I don't know how the battle up on the highway turned out, so I don't know if I should tell you this." Rafaela looked at me, her eyes wide, and her look intense. "One of the last things I remember is Rafael and one of our riflemen moving up on the artillery set up that finally did my squad in. Those were heroic acts that I intend to report on. I wish could tell you if they were successful."

Rafaela's voice quavered. "You were right to tell me. Thank you." Rafaela replied. She was losing control of her emotions. She turned away from me then and left the room.

I resolved to try and get a story about all this to Penny Blue as soon as possible. I needed a typewriter. I hoped that Rafaela could find one I could borrow. I'm sure her operating room calm will return soon, but the constant uncertainty about her cousin Rafael over all these months had to be tough.

PANEL 10 - TREE OF LIFE

Hannah is a crab pot puller on an all woman crab boat working the Bering Sea crab fishery near St. Paul Island, Alaska. In recent weeks marine radio has been reporting increasing hostility in the public exchanges between U.S. President Trump and North Korea President Kim Jong Un. Hannah fears the possibility of Nuclear War in the South. She asks the Captain to let her get off the boat, the "Aleutian Challenger," at St. Paul. She will try to find a pilot to fly her north to her home near Ambler on the Kobuk river above the Arctic Circle.

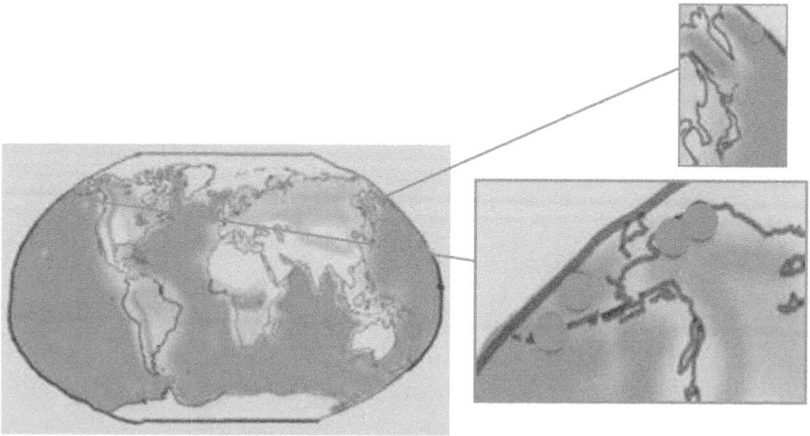

DUTCH HARBOR, ST.PAUL ISLAND, KOTZEBUE, AMBLER WESTWARD, ALASKA. SOMETIME IN THE FUTURE.

PROTAGONIST(S)

Hannah Vitti. - A young Alaska Native woman who is of mixed Inupiaq and Italian parentage.

Mary Trimble – Mother of Hannah, was born to her mother while Mary was an undergraduate student at the University of Alaska – Fairbanks. Mary is Inupiaq born in Point Hope, and grew up in Kotzebue, Alatna, Anaktuvuk Pass, and Fairbanks where she attended school. She graduated from the University with a degree in visual Art. Mary now lives in Ambler in a rough log cabin that she built herself. She makes Northwest Inupiaq art objects for the tourists who come on bus/airplane tours to Anaktuvuk Pass, and for selected tourist outlets in Fairbanks, Anchorage, and Juneau. She spends her days making art whenever possible, after first attending to the needs of her customers around the world who continue to ask her for more of her art work.

Antonio Vitti – Father of Hannah. Athabaskan and Italian, born of an Athabaskan mother in Allakaket on the Koyukuk River, now living in Anchorage.

ANTAGONIST(S) LESSER CHARACTERS

Benny Hanford - An Inupiaq friend of Hannah, born and raised in Kotzebue, who works for a private air charter operator. He flies the company's 14 passenger Cessna Caravan turbo prop aircraft that he keeps in Kotzebue. The plane must be always ready for local charters, and schedule work among Kobuk communities for passenger and freight charter work up and down the Kobuk Valley. Charter and passenger work may take him as far east as Bettles, as far south as Nome, as far north as Barrow and as far west as Point Hope.

Marian Delaney - Captain of the Bering Sea Crab Boat, the "Aleutian Challenger."

Marnie Thompson - Engineer, deck Boss, and second in command to Delaney on the "Aleutian Challenger."

Sadie Smith - Ship's cook, dishwasher, and occasional deckhand for pot setting and recovery when not doing kitchen chores on the "Challenger."

Ann Ruggles - Deck hand, pot puller, Hannah's deck supervisor when setting and recovering pots on the "Challenger."

Inish O'Brien - Second engineer and Crane operator when setting and recovering pots.

David Washington - One of the families of the "White Eskimos" who came to the Kobuk to live off the land in the early 1960s. Now living in Ambler. A friend and mentor to Hannah.

MINOR CHARACTERS

Reinard Saarsgard - (Norwegian. Nick name: Reino) an old man who has been a ham radio geek all his life. After leaving a technical job at ARC when that company was purchased by GCI, he retired to Ambler to further pursue his hobby building radio communication gear.

The most common maps of North America and the 50 states of the United State show the State of Alaska as a small island just off the coast of Santa Catalina, California. The State of Hawaii is often shown south of El Paso, Texas. Many Americans have very little sense of the Arctic lands under the authority and protection of the Sovereign United States of America.

The most distant of these coasts are so far away from the rest of the contiguous 48 states that they are in the eastern hemisphere… closer to China, Russia, and North Korea than to the U.S.

The largest islands in the cluster Attu, Shemya, and Agattu, are located west of the Anti-meridian at 174 degrees East Longitude; roughly 1,000 miles south of the Arctic Circle, and 2750 miles northwest of Seattle. These islands, with Kiska, were invaded by the Japanese in May 1942, months after the bombing of Pearl Harbor the previous December. They were held for a year, then re-taken by allied military forces in June 1943, at a cost of nearly 5,000 allied dead and wounded. The attack on these three distant islands became the first U.S territories invaded, occupied by and re-taken from a hostile war-making foreign government since the War of 1812.

This is a story about the future. In the early fall of a not too distant future year, the North Koreans send a low-yield tactical nuclear weapon into a cluster of small islands at the far western end of Alaska's Aleutian Chain. The aging and under-funded systems and their poorly trained operators on Attu could not – on their own - detect the escalation of the threat from North Korea in the months leading up to the launch, nor could they detect the incoming missile once the button was pushed in Pyongyang.

=====

Captain Marian Delaney of the crab boat Aleutian Challenger is working the crab fishing grounds near St. Paul Island in the Bering Sea. It is October; the skies are clear, but the winds are strong, and the seas have been roughened up with 20 foot waves and white caps. Delaney's load of crab is meagre. The crab population has been on the decline for the past many years, and the prices have declined enough to discourage many of the usual crab fishermen.

The glory days of crab fishing out of Dutch Harbor have mostly passed into history, but Delaney and the boat's owners, Tom and Marjorie Kettleman out of Kent, Washington have decided to keep the crew and the boat in good fishing form, just in case the fishery has improved, perhaps by magic, in this season.

"So far," Delaney mutters to herself, "…there is not much magic in this fishery."

'Hey Cap. Here's some coffee."

Delaney took the cup from Marnie Thompson, her Engineer, and the boat's second in command. When necessary, Marnie could handle the job of deck boss better than anybody else on the deck, including Delaney herself. The deck boss job required a lot of finesse in raising and lowering the heavy crab pots. Marnie's skills and experience are particularly important in swinging the pots around the deck when the deck hands are relatively inexperienced, and the seas are as rough as they are today.

====

It is late October. The crab have stopped moving so the season is nearly finished. Delaney decides to do one more set of 10 pots in an area where she believes there are crab, but she only has a hunch as to how much.

At dinner that evening, while Ann Ruggles maintains the helm, Delaney announces her plan. "I want to do one more small set. I think there are some crab in a little depression on the edge of the shelf about 70 miles southwest of St. Paul. I'd like to do a set of 10 pots before we quit. Once we do that set, we'll head for St Paul to unload and fuel up. Do any of you have any questions or concerns? Any bitches about the trip, or my handling of the boat?"

Inish spoke. "Nah. Yer OK, Cap. Though I wish ye'd got us more crab. My Guinness budget is all shot to hell after this shitty season."

Marnie replied. "I doubt if the seasons ahead look much better; maybe after the crab stocks have had a chance to rebuild."

Sadie the gray haired and slightly overweight cook chimed in. "I'm with you Insh. I'm going to have to get a job in Dutch flippin' burgers this winter. I can't even afford to get to Anchorage to see my buds at Chilkoot Charleys for football and beer."

"I'd like to get off at St. Paul," Hannah said. "I'm worried about the war talk, and my mother is probably worried too. She will need some help if things start to go sideways in the States. If I can use the satellite phone once we are done with the set, I can ask my friend with an airplane to come down from Kotzebue, pick me up in St. Paul, and take me up the Kobuk to my mom's place."

Delaney looked over toward Marnie Thompson. "Do you see any problem with that, Marnie?"

"Nah, Cap. I'm OK with it. I'm getting a little concerned about the war talk myself, but there is no way in hell I'm going to go back to my family in Georgia, even I wanted to. Can I come up the Kobuk with you, Hannah?" "We can talk about it, if you are serious," Hannah said.

"I might be serious. I will think about it."

=====

The pots are big, and they are heavy. The pots can measure 7' by 7' by 3'. They are made out of steel and weigh 200 to 400 lbs. empty. When loaded with bait and enough rope to reach from the sea floor to the surface, a pot can weigh close to 500 lbs. A typical crab boat might carry between 100 and 150 pots. These are all carried on the boat deck, where they are loaded, unloaded, and moved around by a crane permanently fixed to the boat's hull structure. When the crane is moving a 500 lb. crab pot from the stack to the baiting station, then slung to the side and dropped over the side to make the "set", the intricate movements by the deck boss, the pot wrangler, the shot and bait wrangler, and the crane operator form a kind of ballet.

If things go well, and the seas are relatively calm, the time required to set a single pot might be between 2 and 3 minutes. If the seas are rough, the potential dangers to the crew go up exponentially, as does the time required for a set. Loss of life, or serious injury among crab boat crew members has come down over the years largely because of tighter rules about safety equipment, procedures, and training, but each set or recovery, rough seas or calm, are fraught with high risk to the deck crew.

When the time comes to do a "set", each pot must be baited with herring, codfish, or something else a crab might find tasty. Each baited pot must then be dropped overboard with a length of rope long enough to mark the location of the pot on the bottom connected to a float to show the location of the pot on the surface. When preparing for a set the helmsman must move the boat to a designated location, allow enough time to set the pot, until the deck boss calls the length of rope as a number of "shots" to the recorder on the wheel deck. The recorder also notes the longitude and latitude with the pot's Reference number.

The helmsman then moves to the next location perhaps 200 - 400 feet away, then halt and hold the boat so the deck crew can set the next pot. This intense process must go on until the number of pots scheduled for a particular set have been deployed.

Delaney knows that Marnie is fully capable of being the official Captain of the boat, but she still needs to take another Coast Guard certification course and test to get properly licensed. As soon as the season is over in another couple of weeks, she will work with Marnie to help her pass that test. It has often been said that boat crew members in the Bering Sea crab fishery are in the most dangerous job in the world. Over the years since statehood, several members of working boat crews have died in the performance of their jobs, but improvements in safety among other things resulted in no deaths in the Alaska fishery during the 2014 – 15 fishing year.

Unfortunately, in February 2017, all 6 crewmen were lost as their ship, the "Destination," sank in rough seas just north of St. George Island.

In the Bering Sea crab fishery, more than any other single person, the deck boss, is the one person who holds the life, death, and injury of the deck crew in his or her hands during pot set and recovery operations. The biggest problem is moving the heavy pots around on the boat deck, particularly in the heavy seas typical of the peak of October's crab fishing season.

Once set, the pots are allowed to "soak" for a day or two. The soak allows the local crabs on the bottom to smell the bait and migrate to the pots. The crew makes use of the soak time to prepare to retrieve all the pots in the set. After the recovery they hope to have a lot of live, mature crab to put in the tanks of seawater in the boat's hold. The recovery will be even more dangerous than the set because the pots will weigh much more with their load of crab. A successful "set" brings with it a lot of good news for the crew, but always with the dark and lingering possibility, at any moment during recovery, of some sudden bad news because of inattention or misstep by a crew member, or, worst of all, by a mechanical failure.

=====

The seas are rough near St. Paul island, but the skies are clear. Delaney looks around for the best approach to the breakwater and the docks where they can unload their catch. Delaney will also need to take on additional fuel for the return to Dutch Harbor. Marnie is working the deck with the other crew during the docking procedure; Hannah maintains a watch on the radio and radar to keep track of the various ships and their movements around the breakwater and docking area.

Since there are a fair number of boats moving around the Breakwater and docks, Insh remains near the engines during docking. She listens carefully through headphones for Hannah's monotonous, though necessary repetition of the phrase: "All OK on the deck."

Delaney massages the twin throttles on the boat in a way that eventually causes the ship's bumpers to lightly touch the side of the dock. Marnie jumps ashore to catch and secure the forward line tossed by Sadie, and then the back line tossed by Ann Ruggles.

Finally, Delaney announces that the boat is docked and secured. She shuts down the engines and grabs her briefcase full of paperwork for the Fish and Game Department's fish count rep on the dock. Hannah asks if she can use the sat phone for a call to Kotzebue. Delaney waves her hand in approval.

Hannah steps to the back of the wheel house to use the battery-operated sat phone. She decides to try to call her friend with the Cessna 208 Caravan, Benny Hanford. There is no answer at his number, so she tries another friend, David Washington, a son of one of the "white Eskimo" families who came to the Kobuk Valley in the '60s to live off the land. Most of the White Eskimos have now moved on to other things. One of the children of those families, for example, has now grown up and written two best-selling books about his time up on the Kobuk learning how to make his own hunting gear and sleds and how to make his own clothes out of wolf and caribou skins. Eventually he moved to Anchorage to try to learn how to live in the modern world.

Hannah waits to see if David picks up. She tries to contain her excitement when she hears: "Hello?"

"Hi David. This is Hannah. I'm calling from St. Paul, and I am trying to find Benny. Do you know where he is?"

"Jesus, Hannah. Your mom has been wanting to know where you are. This war talk is scaring her."

"Yes. I know, David. I'm scared, too. I'm trying to get home, but the only way I can get there today or tomorrow is if Benny will agree to fly down to St. Paul to pick me up. Do you think he will be willing to do that?"

"I don't know Hannah. Flying is getting to be more of a problem every day because of the war talk. I'm not sure the feds will let anybody take a civilian airplane into the air from Kotzebue. I think I saw Benny on the street a day or so ago, so I'll go look for him. How can I contact you?"

"I'm calling from a sat phone, but you should be able to get a message to me by Marine radio. I'll stay on the boat until I hear from you. If you are going to look for Benny, I'm going to see if I can get a sandwich and a beer somewhere in town. I need to get off the boat for awhile. It is 1:00pm now.

I'll be back here by 3:00pm. Find Benny first, but then please go tell my mom that I am trying to get to her as fast as I can."

"I'll tell her. Don't worry. I'll call you by 3:00pm, and let you know what I find out about Benny."

"If Benny offers to take off right now and head straight to St. Paul, please don't discourage him. OK?"

"Got it." David said as he hung up the phone on his end.

Hannah hung up her phone, then walked off the boat down the gangway, and started walking up the dock toward the small community in the hopes of finding a burger and beer.

No beer and no burger, but the Philippino food served up at the Trident Café – a short walk from the dock - was tasty, varied, and plentiful. Hannah decided to eat quickly, then take a short walk around the town, possibly up by the Russian Orthodox church. She would sit in the church for awhile in the hopes of calming her nerves before she had to talk to David again. If David can't find Benny, Hannah won't be able to fly north today. She will have to decide whether she needs to get back on the boat and go back to Dutch with them tonight or catch the daily Alaska Airlines flight out St. Paul to Anchorage tomorrow. That makes it a two or three day trip through Anchorage to get back to Kotzebue. Once in Kotzebue, she will still have to find a ride up the Kobuk to Ambler.

Hannah needs to avoid spending all that extra time getting to Ambler. She hopes that David will be able to make contact with Benny, and that Benny can come down this afternoon to pick her up. If she can get back up to Ambler tonight, it will make everything much easier.

====

"Hannah. This is David. I got ahold of Benny. You're in luck. He has been discussing a charter to fly down to St. Paul and St. George to pick up some of the field guys working with the Utility company down in St. George. He will bring everybody back to Kotzebue. The military is involved, and it looks like they will approve the flight because they want those technical skills available in Kotzebue while this war threat is going on. He's got room for you if you can travel light and be ready on the runway with the other two St. Paul guys when he comes in to pick you up."

On hearing David's words Hannah suddenly finds herself shaking in anticipation. She was sitting in the wheel house in the Challenger with Delaney, Insh, and Marnie when the call came in. She needed to take a moment to calm down before speaking. "I'll make sure I'm ready and standing at whatever time and in whichever place Benny wants me to stand."

Hannah looked at Marnie, then paused in her conversation with David. Marnie's eyes were downcast. "I hate to ask, David, but one of my shipmates wants to come with me to Kotz. Do you think there is any chance that Benny would allow that?"

"I don't know, Hannah. It sounds like he may have a full passenger load already. I think I can still talk to him by cell phone. I'll ask him to set a course for St. George first, then come back up to St. Paul. You should be able to contact him with your own cell phone in about 3 hours, so you can ask him yourself. I know he will have to top off his fuel tanks for the run to Kotz, so you should plan to meet him at the aircraft fueling pumps at the south end of the airfield.

"If it is possible to shift stuff around in order to fit the both of you onto the plane that will be the place to do it. Once again, plan to travel with nothing more than a tooth brush. Some of those tech guys may have specialized tools that will have to stay on the plane. I know you can deal with it, but you might both have to sit in the laps of some horny guys for several hours during flight. Can your friend handle that?"

Hannah looked at Marnie, who looked Hannah in the eye, in turn, and nodded yes. "She can handle it, David. So can I."

"OK. I'll try to connect with Benny about his route, and about your traveling friend. If there are any initial hang ups from Benny, I'll call you again on the sat phone.

"Make sure your phone is charged up. It will also be good to have a charged spare handy. You may have to do some negotiating with Benny and, probably, with some of the passengers to make this all work."

"Thank you for all your help, David. I'll call you on the sat phone as soon as Benny and I have worked out all the details of what will happen when he lands here. Either Benny or I will contact you from the air as soon as we can connect with ground communication in Kotz. "Thank you very much, David. You have been a huge help. I owe you so much.

"I want to see you safely home," David said.

"Please let my mom know what is happening." Hannah rang off and closed the sat phone circuit.

====

The sun had been below the distant horizon for a few hours leaving a fading and shrinking twilight glow by the time Benny was able to make contact with Hannah's cell phone. "I'm about an hour out of St. George, Hannah. David passed your message along to me. I will do everything I can to get you and your friend back to Kotzebue tonight, but it may require some work and some discomfort on your part. At this point I don't know who my passengers will be, or how many there are. The military is paying for this flight, and they are handling who sits in which seats and how many pounds of freight I will be carrying.

"They know what you want to do, and they said they will do everything they can on their end to make it work. I'm going to ring off now, though, so I can concentrate on St. George. As soon as I am airborne out of there, I'll contact you again. Make sure your cell phone batteries are charged. I'm glad you are OK."

"David says I can't take anything more than a toothbrush. Is that still true?"

"Be prepared to travel with as little as possible. You can bring some kind of a travel bag with you to the airport, but if I say you have leave it on the ground, you can't argue with me. Ok? I'm sorry to be such a hard ass."

"It's Ok, Benny. I understand. I'll talk to you soon."

"One more thing, Hannah. If I tell say you can go, but your friend has to stay on the ground, you can't argue with me about that either. Am I clear?"

"Yes. Benny. That is very clear. I'll make sure she understands. Her name is Marnie, by the way." Hannah closed the cell phone link. The entire crew of the Challenger was now gathered in the wheel house.

"Did you get that, Marnie?" Hannah asked.

"Yes. I am clear on that. Thanks for doing all you can to help me."

Hannah turned to Captain Delaney. "Well, Cap. In a way whether Marnie comes with me or goes back to Dutch with you depends on when you want to get underway. Do you know, yet?"

"I can wait for some resolution of this as long as it happens within the next couple of hours. I think there is some Siberian weather moving in tomorrow, so I will want to get underway before midnight tonight.

You and Benny will need to think about beating that storm, too. Stay alert, and keep things moving."

Hannah spoke. "Marnie and I need to figure out what we are doing so we'll come back up as soon as we are done in an hour or so. Then it will be a matter of waiting for Benny's call. Can one of you guys see if you can line us up with a ride to the airport… and back if necessary?"

Insh got up to get a cup of coffee. "I'll make sure you have transportation, sweetie. Don't worry."

Benny called Hannah about two hours later, around 11:30. I'll be on the ground in about 30 minutes, Hannah. Can you guys meet me at the fuel pumps at the airport?"

"Yes. How much space do you have?" Hannah asked.

"I've got one seat with me in the cockpit, and room under it for one travel bag. There are 10 passenger seats behind the cockpit. They are all men of various ages. One of you may have to sit on the other's lap, but I can't have you doing that in the cockpit. I may have to move one of the guys into the cockpit seat, and you will have to share a seat in the back. Will that work for you?

I need to warn you that we have weather coming in from the west. It's already pretty bumpy. I've got one guy freaking out in the back. I've warned him that the weather may get worse as we get closer to Kotzebue, but If he wants to stay on the plane, you and your friend – Marnie; …is that her name? - may have to stay strapped in a single seat." "We'll make it work." Hannah replied.

The crew on the boat assured both Hannah and Marnie that they would have bunks on the Challenger whenever they wanted to return to work. Hugs and kisses, even a few tears, were shared all around, and the women were on their way. The airport was about 2 and half miles from the dock. Insh had borrowed a jeep from a security guard at the fish processing plant. He had been walking the dock on swing shift. "Sadie gave me some cookies to help convince the security guard that loaning his jeep to me for this worthy cause is a very human thing to do; … that plus showing some décolletage had the effect I was looking for." Insh gave the two travelers a lewd wink and a big smile as she pulled the Jeep away from the Challenger.

Twenty minutes later the runway lights came on, lighting up the entire area around the airfield. A couple of minutes after that, the women could hear Benny's plane approaching for a landing to the north. Then they saw his landing lights come on. Westerly winds across the runway seemed squirrely, but not too strong yet. Even so, the wings dipped up and down in response to sudden gusts. Benny brought the plane down smoothly, first the left rear wheel then the right, then the nose. The touchdown was followed by a reduction in the whine and a change in pitch from the turboprop engine and prop as Benny managed the controls to bring the aircraft to a stop in as short a space as possible. He had to manage the tricky crosswind landing in a way that makes it possible to take off again if he had to abort, while making sure that a strong gust of wind over the airfoil did not accidently cause the plane to rise uncontrolled from the runway after touching down.

Hannah spoke into her cell phone while Benny taxied the plane to the pumps. "Nice landing, Benny. I am so glad you are able to get here."

"Hi, Hannah. Nice to know you are waiting for me. I think you may be in luck. I have to talk to my freaked-out passenger about his options if he doesn't want to continue the flight, but I may have two seats for you and Marnie."

Hannah responds. "If he is that worried about flying in this weather, maybe I can persuade my Captain to take him on the boat when she takes it back to Dutch this evening. I'll introduce him to Insh – who brought us here in a borrowed Jeep – if you want me to offer that suggestion."

There was a pause as Benny considered this. "I'll discuss it with him. Just be ready to grab your seat – or seats - and get strapped in as soon as I finish fueling. I want to do everything I can to beat this storm. We will all talk further in a couple of minutes."

Hannah turned to Marnie and Insh. "What do you guys think of me offering that guy a place on the boat when you all leave for Dutch?"

"I don't have a problem," Insh said, "but why don't you call the Cap now and find out for yourself before you talk to this guy."

"Good idea. Do you have any thoughts on this, Marnie?"

"No. I have thought a lot about this, and I feel like you guys are doing so much to help me that if I choose not to go, I will be letting you down. By now, though, I've decided that I will do almost anything – including hiking into the Alaska bush somewhere above the Arctic Circle – to avoid trying to get back to my parents farm in Georgia. Unless your pilot friend shuts and locks the door in my face, I'm in all the way with you."

Hannah gave Marnie a quick hug. "Good. Your resolve will be important in this."

"Hi, Cap. It's Hannah. One of the passengers on Benny's plane has got a bad case of the Jitters over this storm. The bottom line is this: Can I offer this guy some space on the Challenger so he can get to Dutch with you and the others? That way he can catch the morning flight to Anchorage, and possibly a same day flight from Anchorage to Kotz. If he goes with you, Marnie and I will both have a seat on Benny's plane.

"Otherwise it will probably be a very uncomfortable flight for both of us."

"What do Marnie and Insh think about this?"

"I think Marnie will ride in the baggage compartment if necessary. Insh says she has no problem with the idea of a male passenger on the boat. "

Captain Delaney had more questions. "Have you met this guy yet?"

"No. They are still taxiing up to the pumps."

"My main concern is that he might be drunk, or drugged up, or he might have a gun and some kind of hateful attitude toward women, or some kind of combination of all of these. I can't allow anybody like that on my boat; …unless he is so drunk that he will pass out and stay passed out all the way to Dutch.

"I'm sure you are in a hurry to get this resolved and get your flight underway, but you are going to have to make an assessment of this guy's condition and be ready to assure me that he won't cause me or any of my crew any problems at all. If he looks and sounds like he is not drunk or drugged, and he is willing to let me lock up his baggage, I'll probably be OK with him coming aboard.

"Are we clear on that?"

"Yes. Yes we are, Cap. Benny is just now pulling up to the pumps so we should have an answer on this in a minute of two. I'll call you back." Benny brought the plane up to the pumps and shut the engine down. He jumped out of the hatch next to the pilot's seat. He went to the back of the plane and opened the back hatch to the passenger compartment. A guy got out who looked a little wobbly, but otherwise alert and looking around. Benny pointed toward Hannah and the other women. Then he grabbed a step-ladder to begin the fueling process. He yelled toward Hannah: "I'm only on the ground for about 10 minutes, so hurry up and get your business done."

Hannah stepped forward to introduce herself, Marnie, and Insh to the man. "I'm Hannah. I understand you want to get off here."

The man answers. "Yes. I am very much afraid of flying even in the best of circumstances. When this trip came up, I wasn't really given an option to say no. I did grab a bottle of whiskey to calm my nerves, but the bumpy flight from St. George caused me to throw it all up. The pilot said you might have a better way for me to get to Kotzebue without losing too much time getting there."

"Yes. I think so. I am a member of the deck crew on a crab fishing boat. The boat is tied up down at the docks here, waiting for the three of us to work out a way for two of us to get to Kotz tonight. The Captain on the boat is named Marian Delaney. As soon as we are all ready, she will depart with a five woman crew for a fast trip to Dutch Harbor. You will probably be there by morning in time to catch an Alaska Airlines flight to Anchorage. You should be able to catch a daily flight to Kotz from there.

"But it is up to Captain Delaney to decide whether you can have some space on the boat. I can connect you with her by phone. That may be the best and fastest way to work this out."

"I am willing to do whatever the Captain asks in exchange for a ride with you guys."

Insh spoke. "Do you have any booze or drugs in your luggage? Any guns?"

The man answered. "No booze or drugs. I'm the biggest straight arrow you've ever seen except when I fly or have to think about a departing flight that I have to be on. I do have a pistol, a 38 caliber revolver – it is Alaska after all – but I will happily give it over to the Captain during the trip if she allows me on."

Insh turned to Hannah. "He sounds good to me. Go ahead and get the Cap on the phone. I'll give her a briefing and then she and he can work out any final details."

"While you guys do that, I'll go over to talk to Benny about where he wants us to sit." Hannah called Delaney, spoke briefly, then handed the phone to Insh.

She approached Benny on the ladder fueling one of the wing tanks. "I think we are going to be able to work it out with your passenger, so tell us where you want us to sit on the plane"

"Well, the main thing is that I will want you to sit in the co-pilot's seat. It is not really a co-pilot's seat because there is no aileron control there unless I pass mine over to you. But you know that from other flights in this airplane. Didn't my other pilot, Timmy, throw the aileron control over to you once."

Hannah nodded. "Yes. He did."

"Timmy told me You are a cool hand in the cockpit, and I may need some help on this flight. I am fully loaded, and the weather is worsening.

"I've told the passengers who are still on the plane that I don't want any drinking or drug use on this flight. I told them they need to stay alert and awake and they need to monitor the progress of the flight. If they get jumpy, I'll be depending on you and Marnie to calm them down. Are we agreed on all that?"

Benny did not wait for a reply. "What about Marnie, by the way. If she has never experienced this kind of flight situation before she might freak at the first bad bump. What do you think?"

"In my experience on a rolling crab boat, with Marnie as deck boss supervising the loading and unloading of 500 pound crab pots that are swinging around, I would do everything she told me to do. I have absolute faith in her ability and her judgement."

"Good. I'm glad to hear it. There is a seat in the back. Have Marnie take it. I want her to stay alert to the moods of the passengers. If she sees one of them getting antsy, she needs to get up and try to calm them down. Tell her everything about the flight and what I expect of her as I have told it to you. Got it?"

"Got it." Hannah replied. She went back to retrieve Marnie and check on Insh and the sickly passenger. "Have you guys got it worked out with Cap?" "Yup, sweetie pie. Everything is fine. Give me a hug and I'll be on my way with Tom, here," Insh said.

Tom looked up at Hannah, but it was not possible for her to see the pallor on his face, or his sickly expression. "Good luck, Tom. I hope you can make it to Kotz tomorrow so your boss doesn't get pissed at you. I'm glad this all worked out." They shook hands.

The three women from the Challenger got into a group hug. They danced and whooped for a few moments. They broke up then. Hannah and Marnie to the plane; Insh and "Tom" to the Jeep.

=====

The flight got off the ground to the north at 1:30 am. Benny put the plane into a steady 770 feet per minute climb to its maximum operating altitude at 25,000 feet. He turned the autopilot on. There were some bumps, but nothing scary. Benny hoped to find smooth air at altitude. He flicked a button to drop the oxygen masks down from the ceilings and asked Hannah to make sure every passenger knew how to put the mask on. Most of the passengers seemed to be sleeping. Hannah had to wake them up to help them with their masks. She motioned to Marnie to work her way from the back to the front to help the passengers behind the cockpit. Marnie seemed grateful to have something useful to do.

As Benny turned to pick up his navigation signal Hannah looked back to the west and noticed the City of St. Paul like a ring of bright diamonds floating against the smothering black velvet of night. It reminded her of a scene in one of the boat's lunchroom DVD movies. It was a scene based on a book by Solzhenitsyn. "One Day in the Life of Ivan Denisovich", she thinks.

While looking west, Hannah thought she saw a flash of light far to the southwest, but the light came and went very quickly. It did not linger in the clouds and she could not be sure of what she saw.

======

Around that same time Captain Delaney was pulling away from the dock at St. Paul. As she turned to the west to leave the breakwater she thought she saw a distant flash of light. There was no one else in the wheel house, and it would be another 30 minutes before Sadie would bring up some coffee and then take the wheel while Delaney caught some shut-eye on the cot at the back of the wheelhouse.

Delaney figured she should note the time, direction, longitude, and approximate latitude, of the flash. By plotting direction and compass declination on a sea chart she determined that the flash seemed to be in the direction of Shemya and Attu. Captain Delaney reported these findings to the boat's owners in Seattle via Satellite phone along with the routine report of the unloading and pricing of the load of crab they left in St. Paul.

The owners, Tom and Marjorie Kettleman of Kent, Washington, did not ask about the flash at first. Marjorie did note that the language between US President Trump and North Korean President Kim Jong Un on the news was getting more heated and ugly, as though they were building up to some sort of climax. In light of these increasing tensions, she told Captain Delaney to get the boat to Dutch Harbor and await further instructions. Delaney told her of Hannah's plan to hook a ride with an air charter pilot to try and get back to Ambler to take care of her mom.

Before Delaney hung up the phone, she mentioned that Marnie had decided to go with Hannah up to the Kobuk rather than return to Georgia.

Marjorie Kettleman was on the phone with Delaney but her husband, the Challenger's co-owner, was also listening in. "Well, I'm sorry to lose Marnie from the crew, Captain. Is there anything you could have done to keep her on the boat?"

"I don't think so. We are done for the season, and she had nothing waiting for her in Dutch. I have to give Hannah a lot of credit; she is very focused on what she is doing, and I think Marnie needed that kind of focus to help her sort things out in her own life. If we can avoid war, I am confident that we will see them both back with us next season."

"Truthfully, I don't know if we can avoid war. These little tin pot dictators are going to spit on each other long enough that one of them will soon rub some shit in the face of the other, and guns will start blazing and bombs will start falling among the soldiers and other innocents on both sides of the front lines. I'm hoping there is an adult in the room with Trump, and I hope they hide the nuclear button from him.

Marjorie went on. "Did I hear in your report that you saw a flash of light. Is there anything more to it than that? Is there any lightning in the area? Is it possible that one of the volcanos in the Aleutian Chain decided to let off some steam?"

Delaney responded. "I think any of those are possibilities. For now though, I don't have enough information to know anything. I'll check the radio traffic and see what I can find out. There is a storm forecast from the west, but I haven't checked on it. If I find anything, I'll let you know. Will you be up?"

"Probably not, but don't let that stop you from calling. I'll get up to take your call."

Both women cut their phone links. Delaney moved over to the radio rack, put some headphones on, and began trying to locate stations transmitting news and weather information. After several minutes of finding nothing but static on the usual news and weather information channels Delaney began to wonder what was going on. She was able to find an automatic direction finding signal – called an ADF - associated with a regular radio station in St. George, but, at this time of night, it only played music tapes

An hour later Delaney thought she heard a pop and a dull rumble coming from the west, barely loud enough to be heard above the engine noise, but it did not last long.

Shortly after, Benny and Hannah both felt what felt like an abrupt bump of turbulence, but it did not last long. Benny continued to cruise at 25,000 feet, above most of the weather. There was enough of a moon to cast light and shadows on the cloud tops.

====

Benny was about an hour out from Kotz. He was having trouble raising anybody on the ground in Kotz. He needed the approach control and runway lights turned on, but there was no answer from the tower. He switched to the military channel that he had been issued to him for this flight because of the importance of his passengers.

"Kotzebue tower, this is Cessna N393Charlie Foxtrot coming in with military cargo from St. George and St. Paul. Over?" The silence on the other end was deafening. Benny repeated the call with more vigor. "Kotzebue tower, this is Cessna N393 Charlie Foxtrot coming in with important military cargo from St. George and St. Paul. Is anybody down there? Please respond. I will be descending for approach to the outer marker for runway two seven in about 30 minutes. Please respond. I need somebody to turn the lights on. Over?"

"Hannah. Take the radio mike and repeat the message I just gave them until somebody answers. I need to check some things in case everything is all fucked up on the ground, and I have to do some emergency procedures.

Stay on the military channel, but if you are not having any luck, try some of the other channels. Try your friend David in Ambler on your cell phone. He may be monitoring your number. Try any number you can think of. We are running out of time and fuel, and I need to know what is happening on the ground."

Hannah repeated the call to the military twice more with no success. She picked up her cell phone to make a call to David and was surprised to see that he had left her a text message a few minutes before. She texted back. "David. Thank god. We are approaching Kotzebue but there is no answer on the radio, do you know what is happening?"

Hannah breathed a sigh of relief as David texted back. "Hannah. Try calling me. I think we can still talk, but I am not sure for how long. Call me back."

David came on the line when Hannah called. "I am going to speak fast, Hannah. The military has taken control of all local communications and they have shut down the airport at Kotzebue to any civilian traffic. There was some kind of a bomb, or more likely a missile that landed on Attu at the end of the Aleutian Chain a couple of hours ago. There is a military outpost out there that has gone totally silent.

"I'm able to talk to you because the mining operation at Bornite has its own communication link that doesn't depend on the usual civilian channels. I have access to their system because I do consulting for them. The military's reach is long, so I don't know how long the Bornite link will operate if they want to put a tight lid on communications. You need to listen carefully to everything I say. You need to divert your flight to the field at Ambler.

Electricity in Ambler is generated by back up diesel power locally, and we can light the runway when you get within range.

"A friend of mine in Ambler is a radio geek. We think we can put out an ADF direction signal on a regular radio frequency that your pilot can use to get through the clouds. Once he gets through the clouds on the ADF signal we will have the runway lit, and he can make a normal VFR landing under visual flight rules. Have you got all that?"

"I think so, David, but I want you to talk to Benny. I don't want to screw up the pilot talk. He needs to hear that directly from you."

Hannah turned to Benny. Benny was running instrument checks, but he could see from Hannah's facial expressions and hand gestures that the phone call was important. Hannah handed the phone over.

"Hello David. What's up?"

David repeated all that he had told Hannah. Benny asked that the radio setup in Ambler play some Johnny Cash, so he would know that the ADF signal was specifically for him. "'Ring of Fire', one of my favorites, will be the best," he said.

"One last thing," David said. "The weather is clear below the clouds but there is a weather front coming in from the west and the winds here are squirrely. I'll try to make sure there is a lighted windsock on the field, but I can't guarantee that.

"I think your best approach is runway 1 to the north. You are familiar with the holding patterns and surface problems on that runway, especially in the dark with a cross-wind, but I know you can handle it.

"I'll try to run the grader over the surface but no guarantees on that. Remember that if you have to abort a landing on Runway 1 you need to bear right to avoid the hills beyond the north end of the runway."

"I'm going to head over to the radio setup now to make sure they are on-air, and they have 'Ring of Fire.' On the turn-table. I'll keep this phone connection open as long as I can, but if I lose contact with you, I may not be able to re-connect. Keep your ADF receiver on. We may be able to get a signal up, but we may not have much range; …possibly no more than 25 nautical miles. We can also put up a short wave signal, but I don't know if your plane is equipped to receive it."

"My GPS has not gone down yet, so I still have good lat/long location info coming in. I'm turning now to set a course for dead reckoning to Ambler while I still have GPS. I am showing about 45 minutes to Ambler, by the way. We have short wave receiver capability, but I am not sure I can get a direction out of it except by monitoring signal strength. Do everything you can to direct the beam along magnetic bearing 050/310. We will work on receiving on this end."

"Good luck, Benny. This may be the most important night of your life but no pressure. Put Hannah back on the line. As I come up with new information, she can relay it to you. When you get your plane down on the ground, we can all go get a beer. I'll even buy. Just don't fuck up."
Benny responded. "Thanks, man. If I come in upside down, I'll try to walk away from it. If I walk away, its a good landing, right?"

Hannah turned to the back and motioned to Marnie to come forward. Hannah relayed to Marnie all that she and Benny had become aware of in the past several minutes. Hannah needed Marnie's calm in a crisis to help deal with the passengers.

Benny spoke up. "As soon as you have Marnie up to speed on everything, I want to make a general announcement to the passengers to let them know that we will be diverting to Ambler because of problems on the ground in Kotz. Then I want Marnie to talk to each one of them individually to answer questions, first about why we can't land in Kotz, and why we are set up for a landing in Ambler.

"Once all their questions about landing in Ambler have been answered, then she will need to come back up the aisle to make sure each of them understands emergency procedures if our landing in Ambler does not go well."

Hannah read through the operating manuals as Benny passed them to her. She compared the images of the knobs and gauges in the manuals with the actual knobs and gauges on the instrument panel. She focused most on the radio knobs, buttons, and information read-outs because she figured that keeping the radios going while Benny focused on flying was the most important thing she might have to do.

She worked to be ready to respond to whatever Benny felt he needed from her to bring this difficult and dangerous flight to a successful conclusion. As she noted the instructions Benny was giving her about the function of the various radios and navigation devices she would need to know about, her mind wandered to Benny's life growing up in Kotzebue.

He came from a Whaling family. His father and his grandfather had both been whaling captains. They were in charge of a walrus skin boat about as long as the fuselage of this airplane. When a captain and a whaling crew of 8 to 10 paddlers hunt whales in choppy seas, with waves lapping above the gunwales and into the boat, it was the whaling captain's job to keep everybody calm and focused on the job at hand.

Even on the deck of the "Aleutian Challenger," with the seas running high and the heavy pots swinging around, there was nothing to compare with the dangers of whale hunting far out to sea in an umiak, a small wooden framed boat covered in walrus skins.

The resolution of such dangers is deep in Benny's blood. As she looked at Benny in his baseball cap and headphones, she could see him standing in a parka made out of seal gut, fringed around the face with seal skin and wolverine fur, totally balanced in the prow of a rolling umiak, with a harpoon at the ready, scanning the horizon for the tell-tale venting and rising steam of the whale's exhaled breath.

Whale hunters say that the whales talk among themselves in order to deal with the threat from the hunters, but you can only hear them if you are at peace with yourself in the midst of surrounding dangers. Whale sounds are not heard. They can only be felt as vibrations through the skin of the umiak; a kind of high pitched humming ululation. Benny is like that whaling captain now, Hannah thought.

Benny turned away from his instrument panel to show Hannah the combined compass and ADF signal gauge. He tapped the gauge with his finger and said: "I want you to keep a close eye on this instrument now. We are getting close enough to Ambler that we should soon be able to hear Johnny Cash playing 'Ring of Fire.' That means we can pick up the radio station ADF signal on this gauge if they have been able to put one out. When that double-barred needle jumps, let me know right away."

After a few more minutes, the double-barred needle jumped. Hannah touched Benny on the shoulder and pointed at the gauge as it jumped up and down around the radio signal. Benny adjusts his course to align with the bearing indicated by the ADF signal. He begins a fairly steep descent through the cloud cover over Ambler. The runway comes into view. Continuing his descent Benny finally sets a left wheel down, but the nose continues to drift to the right. Benny knows he has one chance to correct the drift before the nose drifts out of control. If he loses control, the wind will catch the side of the fuselage and yank it crossways to the runway. At the touchdown speed, with this load, of about 80 miles per hour, the plane will probably wrap itself up into a fiery ball and kill or seriously injure everybody aboard.

Benny did not think about any of this. As soon as the nose began to drift, he instinctively jammed the left rudder and began pulling up on the handbrake next to his seat. In a moment the nose came back to the left, and the right wheel came down to kiss the runway. The nose wheel followed. All the passengers, including Hannah, clapped and yelled their approval for this masterful job by a man who, they were convinced, is the very best airplane pilot they had ever known in their entire lives.

=====

For all the fear of signal loss surrounding the approach to the runway at Ambler the actual approach and landing was pretty straight-forward except for the touchdown drama. Benny pulled the plane up to the airport offices, set the parking brake, then shut the engine down. There was silence for a moment or two as Benny and everybody else contemplated their good fortune in getting back on the ground alive. They each thought, too, of their situation now that their need to plan for their future had been restored by Benny's expert pilotage.

David had brought Hannah's mom to the runway. Hannah grabbed her mom and David and hugged them both, hard. None of the three could not help breaking into tears at the reunion. She motioned for Benny to come over and join in. He only hesitated a moment before joining this very private, very intimate moment.

Marnie hung back until Hannah motioned her over so she could meet David and her mom. Hannah's mom expressed great pleasure at meeting one of Hannah's friends. David went over to the military's technical guys to help them figure out where they were, and where bunks were set up at the local lodge.

David had stashed some drinks and sandwich fixings in the bunk house in case anybody was hungry. It was after the season for local hunting and fishing so there was plenty of room in town. They were very grateful that, in addition to still being alive, they will also have a warm place to sleep for the night.

Tomorrow would be a whole new day, and, possibly, a whole new life, and a whole new world to live in.

David got Hannah, her mom, and Marnie into the jeep for the ride to mom's house. She lived a few miles west of town in a log home she had built herself. Once inside, Hannah could not help but feel a great sense of relief. She collapsed into an over stuffed chair near the wood stove where David worked to get a fire going. She invited her mom to come sit on her lap.

====

Her mom spoke. "I have something to tell you, my sweet baby. We are going to leave tomorrow on a hike to Alatna on the Koyukuk river. It is about 200 miles east. We will go up the Kobuk by boat, and over the hills on foot. It will take us about 30 days I think. I have already packed everything we will need."

"Mom. Why in the world do you want to do this; …and why now?" David and Marnie sat mute on a nearby couch.

"I have been thinking about this ever since we elected the man, whose name I will not say, to the Presidency of the U.S. Whatever terrible thing has happened at the end of the Aleutian chain this morning, it is only the first of many things that will now come home to disturb and disrupt all the good parts of our lives. "

"I don't understand, mom. I hope you will explain whatever is going through your mind to come up with these thoughts."

"It is simple my sweet one. There is a dark cloud among the Inuit. It came with those who live only to take our culture, the things we make that are beautiful; …the things that make us the real people. They do not know the people, and they do not know the spirits of the earth. The dark cloud will not go until those who only have values in money and their objects of vanity have gone from us. I think it is time that we go back to our relatives on the Koyukuk at Alatna until all this evil has passed. That is what our ancestors would have done, and this is what we must also do."

"We have two guests, David and Marnie." Hannah gestured toward the two figures on the couch. "Are you thinking they will go with us?"

"As a people we have an obligation to assure the safety, security, and lodging of all our guests. Of course they may come with us, but they themselves must choose to go.

The conversation had gotten too intense. Hannah needed a break from it. The others did too. Hannah sat quietly for a moment then, as a diversion, looked around the walls of her mom's house.

Mary's house is crafted out of rough local logs. Two by fours frame the doors and windows. The roof is made of sod. Even so, the walls contain several beautiful works representing the art of the Inupiaq, the Athabaskan, and other works by members of lesser known tribes in the valley between the Brooks Range in the north, and the Alaska Range in the south. She had many caribou skin masks with many expressions, all made by Nunamiut artists friends of Mary who live in Anaktuvuk Pass.

One of the works on the wall is intriguing to Hannah. She had not seen it before now. It is made from varieties of grasses, twigs, and dried berries, all laid on a background painted to convey an impression of a distant range of mountains shrouded by low clouds. It reminded Hannah of the mountains of the distant darkened Brooks Range with bright streams zig-zagging down the front face of the mountains to connect in a braided outflow into the great Yukon.

The artist had shrouded the berries with some colored lichens in a way that made them appear, in the distance, to be clothed in bright fashions. Some of the twigs the figures held represented weapons and packs, as though they were hunting, or just traveling, or perhaps they were on their way to war.

"Did you do this, mom?" Hannah asked.

"Yes." Mary replied."

"When did you do it? What does it represent?"

Mary looked away. Her lips crinkled in a small smile. "I had a vision in the spring last year. I remember my grandfather telling me about the time when a new group of white people like those who built the sun bombs used to destroy the civilian populations of Hiroshima and Nagasaki in Japan; who put money above the good of the Inupiaq. They came out to our lands near Point Hope. They planned to make huge holes in the earth at what the white man now calls Cape Thompson. They would have used their terribly powerful sun bombs.

These new weapons steal the power of the sun that gives warmth, light, and life and they transform it into an evil and all-consuming fire that eats every living thing. This new group wanted to blow big holes in the ground where the caribou feed. No one knew why, apparently it was for their own strange purposes. They gave no thought to the needs of others who lived there and hunted the caribou and grouse there.

My grandfather gathered his two young daughters and son up with his wife and all went to visit their relatives in Alatna then too. One of the daughters was my mother, Ida Lincoln. That is why I want to go to Alatna now. It is because my grandfather took our family there to keep our family safe from the warped and corrupted sun power that now threatens us again from the south.

"How will we travel," Hannah asked. "I have never been over the trail to Alatna."

"The trail is an ancient one, though not much used these days. There are ancient stories from our family about those who took the trail to escape the many threats that came up the Kobuk river valley as the glaciers melted and the seas rose. Those of our family who made the long journey on foot came to the river we now call the Koyukuk where they established the community of Alatna. In the early days, some crossed the river and continued up to the pass in the Brooks range where they established the community of Anaktuvuk.

"Anaktuvuk is a place of great beauty. It is a magical place where the caribou in their great numbers leave their winter grounds in the mountains headed for their traditional birthing grounds on the Arctic Coastal Plain in the spring. From Anaktuvuk we can look north to the Arctic Ocean, and south to the Yukon river. It is a place of great power, and the Inupiaq feel the constant pull toward that place.

"Eventually, Athabaskan Indians began to come up the Koyukuk with their own interests in the game in the headwaters. When the Athabaskans encountered our people, we fought all their attempts to cross the river. Several terrible battles were fought. Eventually though we found ways to stop the slaughter of our families and our young people; ways that secured and preserved the Inupiaq interests in Alatna, and the Athabaskan interests in Allakaket.

David spoke for the first time. "Have you been over this trail, Mary?"

"No. I have not, but my vision of the journey I must take is clear. I have no doubt that I must go with my daughter Hannah, and I have no doubt that I will find my way. You and Marnie are welcome to join us, as we will need the kind of help that two strong young people can provide."

Marnie now spoke. "There was no room on Benny's airplane, so I couldn't bring any of my winter gear from the 'Challenger.' I have no clothing for being outside for any length of time this time of year, let alone start a hike of several days, even if I wanted to go with you."

Mary responded. "Well, first, I don't expect it will take 30 days because the man who set up the radio signals when you came in on Benny's airplane has a river boat. Reino charters the boat for fishing trips up the Kobuk into the headwaters. He has offered to take as many up the river as he can fit in his river boat. He thinks there may be ice as the elevation rises and the temperatures continue to drop, but he will take us as far as he can.

"But we have to decide, this night, who will go. It is now early fall. We have been lucky, so far, that the river has not yet frozen. Freeze-up is very late this year, probably the effects of the global warming that the people of the sun bomb have been lying about. We need to take advantage of this late freeze because an overnight freeze, if it partially freezes the ground will make river travel much more dangerous. It will compound the many difficulties already ahead of us while we travel overland.

"I don't think clothes will be a problem for you, Marnie, nor for my daughter. I have been gathering a lot of clothing and gear for such a journey for over a year now. I assume that you have plenty of gear available, David, and I assume you will be willing to share as necessary; … that is if you decide to go."

"This is such an amazing night," David said. "I thought my work would be done, here, on this night, as soon as Benny got that overloaded airplane on the ground with all the precious cargo… particularly your daughter Hannah." David thought a moment, then squeezed the arm of Marnie. "And the other precious cargo, this beautiful woman sitting next to me.

"What do you think, Marnie? I have enough gear for the both of us. From what Hannah has told me of your work as a deck boss, I have no doubt that you will be able to deal with whatever inconveniences arise to block our progress on this trek. Do you?" David left his question hanging in the air as Marnie averted her eyes from the group.

She spoke. "This is so sudden, and I have so many things I must think about. But I know there is no time to think about them. Thank you, David, for your expression of confidence in me, and for your expression of affection. And thank you Hannah for letting me come on this homecoming with you; …and Mary; thank you for even thinking about taking a green cheechako from a cracker family in Georgia - who doesn't even have a pair of hiking boots - with you on this journey. It is so important to you and to all of the people here. "

Marnie paused for a moment. "But I am ready to go if you will take me."

"I am too," David said. "Marnie and I can go over to my place to start packing gear. We will see you back here in a couple of hours."

"Mom. You know I have always been ready to go on this trip with you, don't you. Lets start packing." Hannah said.

"Yes. My beautiful daughter, I knew you would be ready to go when I said it is time. Can you go over to Reino's house, David? He is waiting to hear our decision. We will try to be underway by 10:00am tomorrow morning."

"There is one more thing, Mom. I want Benny to come with us. He is a good man; … a strong man from a strong Inupiaq family. Can you go find him, David? I know he won't try to get back to Kotz until he is sure the runway will be open and lit. Is he staying in the lodge with the other passengers?" Hannah asked.

"I'll ask Reino to find him while Marnie and I pack our gear." David replied. "I'll tell Reino to bring Benny here and I'll tell him to say nothing to anybody – including Benny - about why Benny is being asked to come to Mary's house."

Hannah asked her mother: "That's OK, Mom, isn't it? I worked with Benny in the cockpit as we had to try and find the Ambler runway. We were over clouds on dead-reckoning, but he never lost sight of where we were going, mom. This hike, and the times ahead will be difficult. I want someone of Benny's character and strength with me. Do you understand me Mom. Please understand."

Mary walked over to her daughter and hugged her. You are my god-given daughter, Hannah. I worry about losing contact with you, especially to someone you might eventually love more than me." Mary averted her eyes for a moment, "But I can't think of a better person to give your love to than Benny. Bring him with us if he will come."

Benny, when presented with Hannah's proposal, did not hesitate to say yes.

====

With Reino's skillful boatmanship on the Kobuk, they were able to get upriver to a point where the ice finally became unpassable. Mary said it was time to get off the boat and prepare for the remaining 90 mile hike east to Alatna on the Koyukuk.

"If we maintain a decent pace," Mary said, "we should be able to get to Alatna within 15 days. David. You, Benny, and Marnie need to find us some fresh meat and some deadwood for a fire. I want to re-check our gear, eat a good meal and then get a good night's sleep.

Hannah had been talking to Reino back at the place where he had put the group ashore. She turned to the group and spoke loud enough to be heard. "Reino has decided to come with us. He wants a hand pulling his boat up out of the water so the ice won't wreck it when it finally freezes."

"Welcome, Reino." Said Mary when she walked over to lend a hand.

Mary is a tough and seasoned traveler, but after a few days overland on the trail, she called for a stop for the night a little earlier than usual. They pooled their rationed food and enjoyed a good meal around the cooking fire. The group slept in a single tent so they could preserve body heat. Early the next morning, Mary woke up before dawn to start a fire. As she looked up at the sky with the early morning glow in the east telling of the coming dawn, Mary looked for the ancient patterns among the stars.

She looked particularly at the Constellation Cassiopeia, across Polaris, the North Star, from the constellation known as the Big Dipper. At this time of night Cassiopeia was upright in her chair and in all her glory.

In Greek myth, Cassiopeia is the bold queen to Cepheus, King of Ethiopia. Cassiopeia believed herself to be much more beautiful than the sea nymphs that surrounded Poseidon, the God of the waters. Poseidon became offended at her audacity. Eventually, he cast Cassiopeia's beautiful daughter Andromeda into the path of a sea monster where, Poseidon hoped, she would be eaten. The young warrior Perseus intervened in Poseidon's plan when he destroyed the sea monster before he could eat Andromeda.

Mary took solace. Unlike Cassiopeia she had brought her daughter Hannah away from threats by the unnatural people, or others, like Poseidon who live beneath the waters and beyond the real people. She thought about their arrival in Alatna on the Koyukuk river. Allakaket lay across the river, and Anaktuvuk Pass lay farther beyond to the northeast. Soon she would be home.

A large black raven with a white feather on its right wing landed within a few paces of Mary. The raven became known to Mary when seeming to respond to a sound like Bw'-ak. She made the sound, and the raven, as always, would respond by sidling up to her and rubbing the top of his head against Mary's extended forearm. Bw'-ak had accompanied the walkers at a distance through their journey from Ambler. Mary's confidence grew in the belief that he would be there for them for the rest of their journey.

PERSONAL NOTE.

Winter jobs driving school bus and summer jobs on highway and dam site survey crews in Alaska's bush paid my bills in college. In March 1962, U.S. President John F. Kennedy created the United States Peace Corps. Peace Corps Volunteers are, then and now, recruited based on their training, skills, work experience and education. There was no age requirement. Once accepted, volunteers are assigned to underdeveloped third world countries. They work in communities alongside local workers, teachers, and farmers trying to better themselves. In late 1967 I was accepted into the Peace Corps as a surveyor to be assigned to a cadastral survey program then just beginning in the backcountry of Afghanistan. Equipment and training support had been supplied to the Afghan government to run the program. The program was financed by the Agency for International Development (AID).

After taking training in Farsi and some specialized aspects of cadastral surveys at a Peace Corps training camp in Colorado I was assigned to the program's headquarters site in Khandahar. Shortly after, I was assigned to work on a monthly rotation in the field with the program's survey crews, all Afghans.

In the field I and a couple of other volunteers who stuck it out through 90 days of specialized survey training and language training in Farsi would spend a month or so on rotation. From the beginning we decided to split up and go out to the field individually with an Afghan crew.

In my experience working with the Afghan field crews in the distant parts of that troubled country there is one incident that I call a "lensing" event. With a "lens" I can look at a specific time, place, event, and people and see with much more clarity the wider and deeper magic in the place. In telling the fictional short stories here I use a variety of lenses to see the larger context of history.

Every day there is so much news out of the Middle-East, and out of Afghanistan, so the Afghanistan effect didn't affect just me; from the highest diplomat to the lowest Army GI grunts in the field, Afghanistan has affected the fortunes and the families of millions across the U.S. and billions across the planet.

At a personal level, it wasn't all about the work, or the romance of service in a harsh, distant, and storied land. There is my meeting a fellow Peace

Corps Volunteer doing small pox vaccinations with the World Health Organization. After some travel around the country and some spectating at the amazing Afghan national sport, Buzkashi, she became my wife and the mother of my three wonderful children.

Afghanistan has always been an important part of my life. It contributes much to the need I feel to write these stories.

Cadastral surveys, sometimes called land surveys, are used to establish land survey control systems. These are the first steps in breaking up large tracks of lands into smaller tracts known as lots. We start the survey by carrying a specialized survey device called a theodolite up to the highest point of land in the general area to be surveyed. In those days before GPS a theodolite used for land surveys measured angles between some of the prominent stars in the sky in relation to the top of the metal stake called a monument. When the angles are tested and certified by the survey chief, the angle and the time are recorded in a survey notebook. In this way the metal stake becomes a monumented point on the Earth which is unique by reason of its measured and fixed angular relationship to a selection of stars visible from earth.

In the US these surveys end up by setting the survey corners on real properties scheduled for eventual sale and development. They are done in the US by the Federal Bureau of Land Management. Jobs on these kinds of land surveys are usually far from civilization. For guys and girls who work, and who like the feel and experience of mountain tops, wilderness, and rugged territory in general, working on a survey field crew is a great job.

The "Lens" of my narrative here focuses on the second largest city in Afghanistan, Khandahar; my duty station for my work with Afghan survey crews. The city is a continuing flash point in the Afghan / U.S war against the Taliban. It is Khandahar; a city perhaps founded by, and certainly renamed by, Alexander the Great in his march through the area during the winter of 330/329 BCE.

Khandahar is about 300 miles southwest of the capital at Kabul. The road is a narrow, two-lane paved highway. Afghan field survey crews worked their surveys mostly near the trade routes traveled cross-country by camel caravans. The nomadic peoples in these caravans travel with their extended families of infants, elders, and cousins from near and far. With the families are perhaps as many as 200 animals including camels, sheep, burros, and dogs. They were known among the Afghans as "Kochis."

The Kochis and their ancestors have traveled and traded along these trade routes since the beginnings of history. The routes connect their summer trading areas in the north and their winter trading areas in the south. The

Kochis are an independent and colorful people, who have little patience with shopkeepers who often want more than a fair price for their goods.

Kochis are particularly unenthusiastic with those who try to interfere with their traditional movements along their trade routes. They are unwilling to talk about international borders, customs duties and fees (called Baksheesh') when they trade across the western Afghan borders with Iran, Pakistan, and across the northern Afghan border into what used to be the southern border of Russia but are now the southern borders of the independent states of Tajikistan, and Kazakhstan.

In thinking about the Kochis are their fluid view of national borders, I always thought it ironic when I, a U.S Peace Corps volunteer, was invited by my Afghan survey crew mates to go play volleyball with some Russian dam site surveyors working nearby. It was the middle of the cold war, after all. Kochi women do not cover their faces as they walk among the shopkeepers in the bazaars; shopkeepers who may, secretly, consider uncovered faces on Afghan women to be blasphemous. Blasphemous or not, the face-to-face negotiations between Kochi women and local shopkeepers in the bazaars along the Kochi trade routes always bring something new and exciting to the small villages that they pass through.

These villages rarely see anything of change beyond the harsh and extreme seasonal swings in the temperature and weather through the year in this high desert country. The negotiations over price between devout Muslim shopkeepers and Kochi women with their bare arms and faces can be loud. In my time as an advisor, I worked with Afghan crews in the northwest around Herat and in the agricultural areas 100 miles and more south of Kabul. In these southern provinces tribal conflicts between speakers of Pashto and speakers of Farsi are ancient, deep, and occasionally violent. The sight of a few truckloads of bandoliered Afghans shaking their heads from side to side to a beating drum while waving automatic weapons above their heads happened from time to time.

We worked north of the Hindu Kush mountains around Kunduz and Khanabad. In Kunduz the road to the north leads to Tajikistan, in what was then southern Russia; the road to the east rises into Faizabad where lapis lazuli is mined, and then up the 15,000 foot pass over the Wakhan Corridor into China.

My story here has me as far into southwest Afghanistan at the southwest end of the "Dasht-i Margo", or, in English, the "Desert of Death." There is nothing among the windblown dunes in that place even remotely suggestive of greenery or of life of any kind. There is a small border town ahead of us. Looking beyond the town there is a dusty track that leads out of Afghanistan and into Iran.

My lensing event happened in that border town. The town is Chak'hansur. Our Afghan survey boss in Khandahar, (I'll call him Mohammed Aziz), needed to talk with some of Chak'hansur's religious and community leaders about survey business. He and his driver had to drive down by International Travel All – a kind of heavy duty American SUV built in the 1950s - and he invited two of us PCVs, standing nearby, to ride along.

Mohammed did not have to ask twice. We were in the truck almost before he finished his invitation.

An hour or so out of Khandahar, our driver suddenly turned south off the narrow two-lane paved road that connects Kabul, Khandahar, and Herat. He turned where there was no road, and no signs to tell where we were, where we were going, or what direction we should take to get there. The driver simply headed across the desert. Obviously, this would be no ordinary road trip; …but, then again, nothing is very ordinary in Afghanistan.

After another hour traveling over raw desert we began to see what appeared to be the ruins of ancient buildings, their stone walls mostly collapsed and obscured by wind-blown sand. As we drove further the extent of these ruins grew to extend as far as we could see in all directions. We continued south through the ruins for what seemed - in my memory now - like hours.

At the end of our drive south, we arrived at Chak'hansur in the evening, and checked into a local tea house / hotel. Under the dim lights of oil lanterns and the smells of rice and mutton cooking on an open hearth we sat down to eat. The cooking smells somewhat hid the sweet/sour/acidic smell of hashish. We enjoyed a typical Afghan dinner of rice, mutton, tea, unleavened bread with fresh goat butter and jam, and a sweet pudding for desert greatly favored by the Afghans. Eventually, after talk among the Afghans who joined us we went upstairs for sleep.

Sleep did not come for a while, though. The bugs and fleas were so bad in the room and the bedding that we had to move up to the roof. There, in the cool and silent night air under a flood of bright starlight and constellations against the blackened, unpolluted sky, we slept deeply until early dawn. As the sky began to turn light in the east beyond the silent dunes of the Desert of Death, we were awakened by a Muslim singing a morning call to prayer from the top of a nearby minaret. The singing was clear-voiced, and devout. I was not then, and am not now, a religious person, but to hear this at breaking dawn, in this silent place hundreds of miles across deadly deserts from even a hint of civilization was, frankly, an inspiration. It was even - a surprise to me then, and now – a gut-level call to allow my usually suppressed spirituality to run free. Beyond that the call, even for me, momentarily, was an impulse to faith.

In Islam the call to prayer occurs five times a day. At the call a devout Muslim will, wherever he is, unroll his prayer mat, orient it and himself to face Mecca, get down on his knees, sit on his haunches, bend forward with his hands out-stretched to touch his forehead to his mat. Returning upright, his hands open-palmed cross his chest, touching his fingers to his eyes, his face upturned, his eyes closed, talking silently and reverently to his God. This deep obeisance to Allah happens several times in each prayer session.

The prayer session itself is repeated 5 times each day.

In my experience, there is no more physical or powerful expression of faith than this daily devotional act occurring across the Muslim world. Our supervisor, Mohammed, his driver, and our Muslim fellow-travelers on the roof were no exception to this devotion. I and my fellow PCV were awestruck.

I was reminded again of the power of the call to prayer on a trip to Egypt in 2005. There, a friend I was visiting took me up a narrow winding stair to the top of a minaret in the Muslim section of Cairo. At the top we were about 150 feet above the streets. There was only room for the two of us at the very narrow top of the minaret. We could see across Cairo all the way to the Pyramids some 10 miles distant. The mid-afternoon call started quietly in isolated neighborhoods, then grew, blooming like a rose in sunshine after rain until the sound was almost deafening. At the top of our minaret, the aural impact grew overwhelming as the sound wrapped itself around us.

In Chak'hansur the next day, as we met with some of the community's civic and religious leaders, a caravan of several camels crossed through the city on their way north. Several of the men mounted on the camels were armed with automatic weapons. Bandoliers of cartridges crossed their chests.

These were not uncommon images in the south of Afghanistan, though men in the family-oriented trading caravans were not usually so heavily and so visibly armed. Mohammed Aziz told us that heavy automatic arms and cartridge belts are used on most caravans going to or from Iran.

====

Personal Note

The titles of each of the 10 short stories in this book took root in my wanderings, imagination, and the plain curiosity of my early childhood growing up on a farm in the American Midwest. I have been a scribbler and sketcher from the earliest days in my memory. Our library included a "Funk and Wagnalls" Encyclopedia.

The evolution of the title list that became the structure and the names of stories for this book evolved and continued through elementary school; constant wondering and day-dreaming about the sources of ideas and the wonders of language and fantasy and comic books in middle school; and some beer drinking, drag racing, girl-friends and the wonders of sex in high school. None of these activities helped me with my grades, but some classmates and a helpful teacher of English helped me graduate from high school, though I did not also receive a high grade-point average.

As a senior in high school I had been a member of the national Guard for the last several months of the school year. So, almost immediately after graduation I signed up for basic training in Fort Ord, California. I wanted to start my quest to see the world, and to see all the wonderful things I imagined to be in it. That dream didn't last long. A medical problem lingering from an early attack of polio, washed me out of Army Basic Training, and out of any thought of military service. I thought all my dreams of seeing the world were probably gone.

Dejected, I started working as a highway surveyor in Ketchikan, Alaska. One day, in the fall, after a long muddy day on a road construction survey in one of the wettest areas on the planet, I stepped into the supervisor's office, threw my muddy clothes on the floor and stated my belief that there "...had to be a better way to earn a living!" He said "Have you thought about college?" I had never thought about college. Nobody in the family talked about going to college, nor did any of my high school friends. It was a distant kind of imaginary place that had nothing to do with me

But college was a whole new deal. Thanks to the extra efforts of my survey supervisor in helping me navigate paperwork and find a place to take an entrance exam, I was able to start at the University in Fairbanks in January 1961. In college I began to gather focus. I had a new ability in math discovered in my entrance exam, and I gained a new appreciation for books beyond comic books, and for those slippery beasts called words and their flexible uses when we try to write for, or talk, each of us, one to the other.

My travels across Europe, Russia, Asia, and around the planet after college gave me the experience and drive for writing a book. I traveled by car across the U.S.; by freighter across the Atlantic; by hitch-hiking, train and bus across Europe; and by Train and plane across the U.S.S.R. I needed to review the works of the other writers, and visual and film artists that had gained significance over the course of my adult life.

"Walkabout," is the name of a book, followed later by an Australian movie Directed by Nicholas Roeg. The 1959 book and 1972 movie explore the right of passage of a maturing aboriginal boy who is black. A boy on walkabout must travel by himself for many days and weeks living off the land in the harsh desert until he finds his way home as a man. In the movie his rite of passage gets turned upside down when a terrible and disturbing event throws a young and pretty private school girl and her much younger brother together with the aboriginal boy during their common search for water in the dry and dusty desert. As they continue their quest for day-by-day survival the world changes for them.

The presence of the young white girl with the black aboriginal boy in their common quest for survival in the harsh desert changes everything about the worlds they thought they knew.

Along with walkabout, the other major influence on the writings in fiction in this book is myth; particularly the Greek myths with their origins buried in the oral traditions of their story-tellers. Homer may be one of these oral story-tellers, or he may be one of many story tellers cited by whichever writer cast the first heroic stories of the Greek heroes Achilles and Odysseus to writing. Or maybe Homer didn't exist at all.

The Greek myths make my life choices easy; I've decided I want to become a writer and story-teller like Homer; ...or, maybe, on thinking about it, I would really rather become a painter like Basquiat.

Thanks to Hetty Worth Barthel, my long-time accompanist, story critic, and art apprentice for all that she has done to help me keep this book moving; to Jeff Brown at KTOO for giving me the first public venue for the paintings that have guided my story telling; and to visual artist Lindsay Carron for her penetrating comments on some of my stories.

Jerry Smetzer, June 4, 2018

ACKNOWLEDGMENTS

Sources and Methods

The blank sheet of paper.

This writing project began as a late-in-life consideration of all that I had seen and done while traveling around the world - after college and some work to earn travel money - on a solo adventure. I am a movie-goer. Several years later I learned, from a movie of the same name, that Australian aborigines name this kind of curiosity and exploration of new places a walkabout. A young aborigine on walkabout is transforming himself from a child into an adult. He takes a knife, a bow, some arrows, perhaps, and a few other tools and personal items into the Australian outback. There he must learn to survive, alone, for several weeks; ...possibly for several months.

In the movie, a troubled and disturbing confluence of events put an aboriginal youth on walkabout with a young white private school girl and her much younger brother. They were deep in the outback searching a damp spot on the ground in hopes of finding scarce water. They are far from any hint of civilization; far from the supportive structures of human society. They have no common language. Together the three would have to learn to survive their walkabout together in order to get back to the comforts and security of their home communities.

This cinematic walkabout story has persisted in my memory all my adult life. It has become entangled with all that I have read, studied, and practiced in mathematics, history, literature and the arts in college, in working for a living, and in helping to raise the children. A desire grew in me, over time, to capture both the grossness and grittiness of the walkabout experience with the less gritty more abstract ideas gained by reading, working, and living life in places where having a warm place to sleep and warm food to eat each day is never in doubt.

The idea of writing a book about the aboriginal practice of walkabout in some combination with a figure like Cassiopeia from Greek myth began to form over the summer of 2015. At the time I was completing work on a series of 10 oil-on-canvas paintings. I had recruited a professional model and had begun sketching representational poses in 2013. Together we created pencil, and pen and ink sketches of a few hundred portraits and poses. Along with the poses I had also googled and gathered a set of images, portraits and references to incidents and events relevant to periods of history that have become memorable and important to me over time as cultural markers.

Examples of these markers are the visual objects within the 10 oil paintings that are the visual references for these stories.

It was out of those poses, images, and web references that the topical outline and titles for the 10 pieces of short historical fiction presented here became stories.

In thinking about the paintings each of the 10 appears at first to be a kind of collage of objects. Actually, each is closer in concept to a message notepad lying next to a telephone. Over time a telephone pad becomes a palimpsest of layers. Each layer captures and preserves our most thoughtful impulses when we take note of the important moments and information given to us. With such a pad it is only the writing most deeply felt or most immediately consequential that persists on the pad as each page, or layer, is eventually torn off by use and age.

Layering is a tool commonly used in digital drawing, painting, and motion graphics programs on computers. I use brands like Adobe "Photoshop," and Corel "Paint" extensively, along with a variety of cheap mobile "apps" to create multi-colored, multi-layered compositions of digital imagery of the kind used as references for the 10 paintings discussed here.

Layering is also at the heart of a computer's operating system and programs. When a computer is turned on, or when an app is launched, blocks of code containing groups of algorithms are layered onto the computer's memory in a specific sequence. Taken together all the layers of code in a program are what bring the computer and supported software to life as a useful tool for a writer, or for anyone trying to express himself or herself in a creative way.

One other piece of image manipulation software should be mentioned here. When working with a live model I sometimes found it necessary to visualize a three dimensioned human form in a complex, sometimes animated pose with variable lighting and orientation to a variable point of view. The use of 3D animation software gives me the ability, to look at each frame in a completed animation. With this frame-by-frame analysis I can figure out how to render and light the pose within the two dimensions available on a canvas or computer screen. For those times I use a 3D object manipulation/wire-frame painting and animation program called "Poser."

Using these software tools together in some integrated combination makes it possible to bring all the discrete pieces of digital imagery and information together into a digital composition. When I am ready to paint, this composition guides my application of paint onto canvas in much the same way that a fresco artist applies his design – also known as a cartoon - and paints onto wet plaster.

Each of my completed paintings are a collection of scenes and images representing the period of history suggested in the title of the painting. As with any project involving personal expression in the public space, early promotion is important. When completed, these 10 paintings went on display in the hallways of Juneau's public radio station, KTOO, for four months beginning in December 2015.

The Protagonist.

The central figure in each painting is a visual representation of an intelligent and attractive young woman. Since each painting tells a story, each young woman featured most prominently in a particular panel becomes the story's protagonist. This choice of women protagonists reflects my life-long respect and appreciation for the strong, intelligent women in the arts, and among the friends and family in my life and in my work place experience. I had not thought of fiction in my original concept for this project. It began and continued for a few years as a project in the visual arts as I developed the 10 20" by 26" oil on canvas paintings. At first my plan would have me writing a paragraph of non-fiction text describing some of the historic, social, cultural, and political forces at work as suggested in the historical context of the paintings. I had not realized how inadequate the descriptions would be in giving force to the descriptive narrative.

Then my daughter, Megan, a professional art historian specializing in the art of Pacific Northwest Coast Indians, said she wanted to know more about the women in each of the paintings. "Who is she?" my daughter asked. "Does she have a family; why is she in these stories?"

Since that day, I have spent most of my writing time trying to answer those questions. I resolved to write pieces of short fiction that link - to the best of my ability though I am no historian - the historical context in the published record with the day-to-day living experiences of my fictional protagonist. Her daily life, of course, takes place within the fictional and non-fictional people and events that take place around her. I hope by this book that I have made at least a start toward answering my daughter's questions about the young woman in the paintings.

=====

The tools.

Much of my working experience since college has been in information technology. Writing a book of historical fiction means keeping track of the web links, research data, and other information that I would have to gather, verify, keep current, and make easy to find and retrieve. I know how to set up the simple file structures and web page bookmarks that are expandable, integrated and reasonably secure. As part of the original painting exercise I had developed a reference library of several books, images, and Google search bookmarks appropriate to the painting and title in each panel. I added these to my newly created system of computerized file structures, web links, and databases.

So far it has worked well. My reference database now contains over 250 items. It includes books, movies, lectures, and images. Some of these references I have studied closely, others I have glanced at in order to glean a reference detail, and some I know to have been major influences in whatever writing and writing style that I hope to bring to these stories.

The reference list does not include the Googled bookmarks of relevant pages on the internet. The rough page count of web pages referenced probably adds another 200 to the reference count.

In history, maps are important. The tools available in Google Earth are used extensively in the research for each story. Without the visual imagery in the geographic facts of the historic places, these stories in text could not carry the fictional narratives. Unfortunately, Google's copyright policy in using images built with Google Earth is so ambiguous as to make them unusable for publication. Using my other image manipulation tools and royalty-free map projections I have drawn my own maps for the book's cover and for the pages titled "Story Geography and Story People" that open each story.

There is one comment I should make on the use of digital search tools on the internet. They are essential in any kind of creative work – whether fiction or non-fiction - that must be tied to the "facts" of time and place on the public record. Like words, facts are slippery beasts that must be carefully hunted, and rigorously cross-checked against multiple sources. Even then there is no absolute assurance that the "fact" will stay static and impermeable from search to search. There are many reasons for this, but most have to do with the search through digital systems where each branch in the search path depends on a binary yes/no choice. Whichever choice is made, the path through the alternative choice is no longer available in a simple digital search.

Such lost search paths may contain gold mines of data, but if the cross-checking is not sufficiently rigorous they won't be found. A computerized search mechanism that keeps all the paths and their alternatives available to the search algorithm until it is complete is not yet available. That algorithm probably won't be available in any meaningful way until quantum computers come into more common use over the next several years.

Finally, I have become well experienced in the use of the expensive industry standard for image manipulation, Adobe Photoshop; and for words and data, Microsoft Office. Photoshop became essential in designing and assembling the hundreds of images layered into each of the compositions for the story panels. I did not usually dip my brush in oil paint until I had a digital photoshopped composition file that gave visual hints of everything I wanted to say. Such compositions should also give guidance in the process of expressing complex ideas in the 2-dimensioned visual space of a piece of canvas or on the visual matrix in an electronic digital screen.

Setting these computerized structures up and loading the data and information I had already used for the paintings took a bit of time. However, the savings in time these data and information structures make possible when looking for specific references while composing either text or images makes the time invested in design, setup, and data loading a significant factor in any time or budget estimate.

Searching for lost references.

The success of any kind of complex writing project depends on capturing the muse when she overcomes her shyness or reluctance enough to approach the writer. When she is nearby a writer must capture the moment by writing and writing, then writing some more. Any disruption that takes away from her presence at your shoulder will break the mood, the focus, and the energy. She will fly away and her flight will destroy the writer's ability to write until she returns.

I knew this writing project would depend on having as few disruptions from the writing as possible. The time wasted trying to find specific references that are not where you thought they were is at the top of my list of time-wasting disruptions in the process of writing creative fiction and non-fiction. Properly set up, a computer data system with a system for organizing books and other physical references can eliminate much of that time and waste but they cannot eliminate all of it.

If I learned nothing else from my years working in information technology, I did learn there is no perfect digital system. By the nature of digital systems built on Boolean yes/no choices, there are many built-in failure points. The only digital systems that can function in real-world production environments are those where the error rates that kill productivity are defined and identified as to their likely occurrence. These occurrences are reduced to some agreed upon level where the cost of fixing an error can be assessed against the cost of replacing the hardware or software part that is the source of each of the errors.

===

What is the story line and story arc; what are the scenarios?

The visual imagery in the 10 paintings in discussion here having been set for the context of the history and the protagonist; the impulse to create an identity for the protagonist – following up on the question "…who is this woman?" This is the business of writing story lines, story arcs, or scenarios. That writing now begins. These story guides help extend each painting into the fictional characters and events that occupy that piece of history under the "lens."

The terms story lines, story arcs, and scenarios may suggest something similar about story structure, but there are some subtle and important differences; If the writer has not yet moved beyond the blank sheet of paper that sits in front of him or her, he or she may need a story line. A story line, in the form of a couple of sentences, sets the initial time, place, and principle character in the story. It should give the reader at least a glimpse of what kinds of things might happen as the story unfolds.

If the writer knows where the story starts but not where the story ends, he or she may need a story arc. In other words, from the beginning how does the story unfold? Does the protagonist take a trip, decide to stay at home, get caught in a snow storm without a coat? How do these events lead to the final scenes in the story, and how do any conflicts or threads in the early scenes get resolved in the final ones? These are questions that are dealt with in a story arc.

Most stories are made up of several scenes, or events, or incidents that show how the characters interact in various settings and events. Each of these can be described in a scenario that describes emerging conflicts or interactions among the characters that may be introduced.

=====

What is the Writer's Writing Style?

A writer knows that he or she must, sooner or later, have a recognizable writing "style." A style gives the writer identity in the mind of their readers as well to himself or herself. In the simplest possible concept a writing style is the narrative platform created by a writer as he or she refines and adjusts it to tell particular stories. As the writer's oeuvre expands the work gets more complex. Even a beginning writer may find it pays to develop a written style sheet to use as a handy reference and reminder of the writer's preferred way to tell stories, if for no other reason than to keep track of the correct and consistent spelling of character names.

A style sheet is a combination of many elements. The style manual published by the University of Chicago in 1993 is over 900 pages. Rather than try to catalog all of the possible elements of a style before beginning this writing, I felt my time would be better spent by reading books of fiction by authors whose works I have enjoyed and whose works have been recognized for their popularity or their importance to literature over time.

I resolved to read several of the books in English that have defined the diverse and deep scope of English language literature.

Don Quixote by Miguel Cervantes was a good place to start, I thought, though I soon realized that reading anything in translation was not going to give me access to the actual writing style of the writer because I was reading the translator's style. From then on I concentrated my readings on writers who write primarily in English; with the exception, of course, of *Ficciones* by Borges.

My readings continued through many works. *Middlemarch*, by George Elliot, a woman, took me over 800 pages of dense text deep into the politics and social and political nitty-gritty of a small English town in the early 19th century. Another English woman writer, Charlotte Bronte, captured my attention with *Jane Eyre*, a much more readable volume made even more accessible by several movies by different directors and actors.

Writers known for their extreme styles also got their share of my attention. James Joyce's *Ulysses* had sat on my shelf, unread, for years. I finally resolved to read it, and I did read it. I have still not been able to build the courage and commitment needed to read all of Joyce's *Finnegan's Wake.*

Though I have a BA degree in English, the American Nobel Prize winner, William Faulkner, had never had more than a glance from me in college. However, by the time I was a couple of months into my readings, I had read all of his major works; *Light in August*; *Absalom, Absalom*; (a book almost impossible to read, but I read it); *The Sound and the Fury*; *As I Lay Dying*; *Sanctuary*. I read them all, but decided, at the end, that Faulkner's style is a little too personal to work for me as a style.

I read popular contemporary novels such as those written by my favorite mystery writers, James Lee Burke, creator of the magnificent, struggling to stay sober, ex-marine and Vietnam vet; occasionally overwhelmed by the horrors of war; washout from the New Orleans Police Department, Dave Robicheaux, and his big, overweight, violence-prone, pork-pie hatted, pink Cadillac driving, bail-skip-chasing, too often boozed up side kick Cletus.

Elmore Leonard's funky dialogue also came in for a lot of close scrutiny. I enjoy all the work by Elmore that I read, but his stories pale slightly when held against the rich tapestry of the Louisiana bayou around New Iberia, especially down by the bait shop when owner, ex-New Orleans police officer Dave Robicheax, asks his old black retainer and boat, fish, and beer wrangler, Batist, to rig a skiff for a customer; all while the bream jump and splash in the warm summer waters near the dock on Bayou Teche, while the heron try to snatch them, and the three-legged raccoon, Tripod, scampers about looking for bits of oyster in discarded shells.

Many years ago, at a week-long art festival while I was an undergrad at the University of Alaska – Fairbanks, I had the great pleasure of meeting and hanging out with Norman Mailer and Ralph Ellison at a week-long Arts Festival on the campus. This was in the mid 1960s. I had read Mailer's *Naked and the Dead* back then, but never Ellison's *Invisible Man*. Now I read Ellison's book as part of my research into writing styles for this collection of stories.

These two authors could not have been more different in their personal styles, though their best-selling books were, mostly, within the limits, then, of barely acceptable good tastes in mainstream publishing. During their visit Mailer, a New York City ex-combat infantryman and Jew built in the fireplug shape of a prize-fighter, grabbed the male host at one of the artsy, rowdy parties - held in honor of the two best-selling writers - in a big bear hug. The two men proceeded to dance, hugging, around the floor while both uttered loud growling bear noises.

All the while the black, tall, thin, and reserved Ellison took drags on his holdered cigarette and looked at something in the ceiling above the noisy get-together; …or so my memory has it.

Like Eudora Welty's 1941 short story, "Why I live at the P.O." these images of contrasting styles have stuck with me for as long as I have thought about the nature of writing and the creation of a robust American literature.

Three of my favorite fiction books in college were *Catch 22*, by Joseph Heller, *Cat's Cradle* and *Slaughterhouse 5*, both by Kurt Vonnegut. I read them all again. I read a lot of Shakespeare because Juneau is a community full of Shakespeare enthusiasts. There was a festival, and they had all got together to read plays from his "First Canon" when a copy was on loan to our State Museum. I and a friend joined them for a reading of *Titus Andronicus*.

I have become a fan of Seattle writer Neal Stephenson primarily because of his *Baroque Trilogy;* though his first novel, *Snow Crash* is what brought him to my attention. Stephenson's *Trilogy* is a set of three 1,000 page books describing, in fiction, the roots of the Enlightenment as seen through the eyes of a band of adventurers and trouble-makers. The story unfolds around the time of Elizabeth 1 and her Scottish successor after her death, James 1. In the course of their various "walkabouts," Sir Isaac Newton takes over the minting of English coinage, and much of the City of London burns to the ground.

These adventurers counted themselves as friends of the members of English royalty as well as friends of the members from the lowest ranks of English society. They counted as friends of Sir Isaac Newton. As Stephenson weaves his stories and his many characters among the events and historic incidents of the tumultuous times between the plague and the industrial age I became very enamored of Stephenson's story-telling practice. *Cryptonomicon* is also a favorite.

Among modern American writers the Hemingway style is the easiest to read, but I found myself rejecting the content of *The Sun Also Rises*, and *A Farewell to Arms* as being too refined and a little too sweet for my taste. Then I read *For Whom the Bell Tolls*, Hemingway's story of an American volunteer who fought with the liberals and republicans in the Spanish Civil War against Franco's Fascists. I came to the realization that the story is close to being one of the finest works of war fiction that I have ever read. Whatever writing style finally evolves for me, I hope it will carry my fiction across to the reader in as clear and as moving a way as did Hemingway's style in this book.

The movie *Chinatown*, directed by Roman Polanski, and released by Paramount in 1974 must be noted here because it is the finest example of a movie-maker's story-telling style that I have seen. Guillermo Del Toro's *Pan's Labyrinth* comes in as a close second.

Finally, another important part of a writer's style is his or her use of story structure. General considerations of structure are not always obvious in reading because structure is usually relevant only to a particular story. Structure is much more apparent – that is, easier to see - in movies. In my case one movie in particular: *Cloud Atlas*, - produced by the Wachowskis and released in 2012 - offered me the structural pattern I needed for my stories. *Cloud Atlas* linked, loosely, six separate and independent stories over a period of about 300 years.

The *Cloud Atlas* story structure seemed a perfect model for my attempt to write 10 stories loosely linked over a period of about 100,000 years.

Of the 250 references in my Concordance to *Cassiopeia's Quest*, 198 are books, and half of those are works of fiction written in English.

====

The origins of my impulse to write a book length collection-of short historic fiction.

The transformation of these ideas into a book began with a series of large oil on canvas paintings or palimpsests of overlaid images. I completed them for a personal show that went up on the walls of Juneau's public media station, KTOO, in December 2015. These paintings reflect my continuing interest in representing the details of human diversity as individuals struggle within the rigid constraints of history and culture.

My professional model worked with me in several sessions over a period of 3 months in the spring of 2013. After a few sessions the poses began to suggest in my mind an intersection between cultural evolution and the form and humanity of our big-brained species: Homo Sapiens. Each painting represents a marker in human progress, and it is from each of these palimpsests that I have taken each of the 10 stories.

The stories themselves are fiction. I am not a historian, but I have worked carefully and diligently in finding and cross-checking the facts and details of history that become the context and details for the daily lives of my fictional characters.

Finally, there is Cassiopeia. In Greek myth, she is the beautiful black Queen of Ethiopia. She bragged about the superiority of her beauty so vigorously that she managed to anger Poseidon, the Greek God of the Oceans. Poseidon punished Cassiopeia's daughter, the even more beautiful Andromeda, by chaining her naked to a rock to be eaten by a Sea Monster. She was saved by Perseus but the saving of Andromeda further angered Poseidon.

Poseidon then punished Cassiopeia by chaining her to her throne; …a throne that cast her upside down in the night sky across the Pole Star from the Big Dipper for half the night. Her quest is to rise above her anger at this insult. She knows from everything in her life and learning that her status is above that of her constant oppressor, Poseidon. She knows her status is equal only to Zeus at the top of the hierarchy of the Greek Gods.

Forever after, even though chained to a throne rotating around the Pole Star, she is dedicated to her belief in that truth. Her belief is a heart that beats beneath her skin in each of my stories.

Juneau, Alaska, June 2018

About the Author.

Jerry Smetzer lives, writes, paints, and hangs out with his long-time gang of friends in Juneau Alaska, USA. He has lived, studied, and worked in Alaska – almost entirely in Juneau and Fairbanks - since the age of 16. He has read deeply and traveled widely all his life. He has kids born and raised in Fairbanks, now scattered with grandchildren around North America. He was pleased to have gotten together in Florida with them and with all the extended family for his Mom's 100th birthday in June of 2018. "I have always wanted to write a great novel. With the help of all my readers maybe we can make it so with *Cassiopeia's Quest.*"

Made in United States
Cleveland, OH
05 November 2026

47669888R00199